T0360460

ROUTLEDGE LIBRARY EDITIONS: URBANIZATION

Volume 5

SOVIET URBANIZATION

SOVIET URBANIZATION

OLGA MEDVEDKOV

Routledge
Taylor & Francis Group

LONDON AND NEW YORK

First published in 1990 by Routledge

This edition first published in 2018
by Routledge
2 Park Square, Milton Park, Abingdon, Oxon OX14 4RN

and by Routledge
605 Third Avenue, New York, NY 10017

Routledge is an imprint of the Taylor & Francis Group, an informa business

British Library Cataloguing in Publication Data
A catalogue record for this book is available from the British Library

ISBN: 978-0-8153-8014-6 (Set)
ISBN: 978-1-351-21390-5 (Set) (ebk)
ISBN: 978-0-8153-8010-8 (Volume 5) (hbk)
ISBN: 978-1-351-21402-5 (Volume 5) (ebk)

Publisher's Note
The publisher has gone to great lengths to ensure the quality of this reprint but points out that some imperfections in the original copies may be apparent.

Disclaimer
The publisher has made every effort to trace copyright holders and would welcome correspondence from those they have been unable to trace.

SOVIET URBANIZATION

Olga Medvedkov

ROUTLEDGE
London & New York

First published 1990 by Routledge
11 New Fetter Lane, London EC4P 4EE

Simultaneously published in the USA and Canada by Routledge,
a division of Routledge, Chapman and Hall, Inc.
29 West 35th Street, New York, NY 10001

Typeset by Leaper & Gard Ltd, Bristol, England
Printed and bound in Great Britain by
Biddles Ltd, Guildford and King's Lynn

British Library Cataloguing in Publication Data

Medvedkov, Olga, 1949–
    Soviet urbanization.
    1. Soviet union.   Urbanization
    I. Title
    307.7′6′0947

    ISBN 0–415–03869–3

Library of Congress Cataloging in Publication Data

Medvedkov, Olga, 1949–
    Soviet urbanization/Olga Medvedkov.
        p.   cm.
    Bibliography: p.
ISBN 0–415–03869–3
    1. Urbanization–Soviet Union.   2. Cities and towns–Soviet Union–
Growth.   3. City planning–Soviet Union.   I. Title.
HT145.S58M44   1989   89–33998
307.76′0947–dc20   CIP

To my family

Our Russian Past
and
American Future

Contents

Figures

Plates

Tables

Introduction

Scope, purposes, objectives

Soviet urbanization is worthy of attention for a number of reasons. First, deep and rapid changes have made a major world super-power out of old Russia, a country of illiterate peasants. Its cities manifest most of the change. The strength of the Soviet Union – its factories, its well-trained labour, and its military-industrial complex – are located in the cities. For a better understanding of the Soviet Union, for doing business or taking political steps, it is useful to learn about its complex urban structure and processes.

Second, the Soviets' way of implementing urbanization is orig-inal. Disapproval of their dictatorial solutions and experiments should not interfere with an objective assessment of their origin-ality. Their successes and failures carry lessons for the future, and not just in Russia.

The third point combines those aspects of the previous two that help to explain the enigma of burning actuality. I mean an internal mechanism of the present-day crisis in Soviet society. Facts and findings about the Soviet urbanization permit exposing the mechanism.

The job of looking into a crystal ball and making predictions does not go very well with the style of data analysis adopted for this book. Alas, the reader will find hardly any entertaining pre-dictions here at all. This book is addressed to students of Russian realities, who know in advance that a research report may be heavy food.

As I complete this book in the summer of 1988, my view about the challenges for Mr Gorbachev is close to the one stated in a recent volume (Lewin, 1988). Stated briefly, Soviet society is crippled by a gap between the complex fabric of its urban life and the archaic practice of its rulers. The rulers insist on sending orders from above into the society, as the Czars did, as if all the nation

1

were still just a network of garrisons. In this book one may learn how much in modern urbanization is beyond the reach of the simplistic planning mentality of the Soviets.

I will show how an accumulation of factors in the urbanization process can slow down or derail efforts of the planners: for example, when Soviet cities suffer from 'urban pathologies', and stagnate because of dated or exceedingly narrow specializations.

A study of urbanization presents opportunities to observe nearly all aspects of a society. However, any discipline-specific study is limited in scope, and certainly my book is no exception. My observation field is 'now' and 'the immediate past', the 1970s and the 1980s. I highlight constraints for the Soviet urban future derived from the substance of the findings rather than from any bias towards the future. My ultimate goal is to present a better understanding of Soviet urbanization, but my specific research objectives are more limited.

I have selected a limited number of questions, problems and processes for an in-depth analysis. These are questions which a geographer may reasonably answer. At the same time each topic is answered so that an interconnected picture may develop to explain current Soviet urban society.

Chapter 1 starts by examining the hierarchy of the Soviet urban system. The concept of hierarchy is central in Soviet society; it has deep roots in Russian civilization and history. I am trying to measure how well all Soviet cities are adjusted to each other, and what is unusual in their ordering by size. I make comparisons with other national urban networks within the family of Warsaw Pact states, to uncover not just the obvious and generic results of the command type economy, but also more subtle characteristics specific to the Soviet Union.

The topic of urban hierarchy permits us to uncover powerful internal tendencies that work spontaneously. I ask the question: How strong is the interdependence between the urban hierarchy and the development of all national territory? When interdependence is proved to be very strong, I examine how much of its strength may be attributed to a consistent course of modifications implemented by command-type planning.

I am not looking at attempts, declarations, or deployment of resources but solely at the imprint present in urban hierarchies. The end point of that enquiry is not reassuring for centralized planning. Random walk features are strong in the time path of the urban hierarchies studied here, suggesting the impotence of centralized planning, or at least its inability to direct a comprehensive organization for all settlements.

Chapter 2 presents functional profiles of all major Soviet cities. This subject is necessary because of the well-known changes introduced into cities by the policy of rapid industrialization. I assess the imprint of industrialization on Soviet urban life; luckily the most detailed data from the 1970 Soviet Census of Population was available for the job. I test a hypothesis about the most likely forces of growth in the cities during 1970–86. It shows that growth was very much in compliance with the geography of the early investments in industrialization. A widening gap appears between this geography of past successes and the presently important location of raw materials and labour. The data on frequencies of particular urban functions permit me also to locate the centres that produce most of another pathology in the Soviet economy – the overloads for Soviet railways.

Chapter 3 compares two processes of growth and development in 221 major Soviet cities. The first process simply adds more of the same already existing industrial specialization of cities. The second (development) process leads to qualitative transformations, based on the technologies of the Information Age. I demonstrate that the transformations are few. By testing hypotheses I also show that it is possible to detect among the conditions in all Soviet economic regions those factors which contribute either to the industrial or the transformational pattern of urban growth and development.

In Chapter 4 the structural properties of the Soviet urban network are compared with the views of most theorists who have advanced fundamental proposals about better organized growth of the network. In searching for such views and in making comparisons I again look at the situation of the Soviet Union in parallel with other nations of the Warsaw Pact. This method permits me to show that Soviets have in their urban network the most random interplay of different principles of organization. This randomness presents more difficulties for those who would make improvements.

The conclusion gives a critical overview of the findings; here also will be found my personal views about how the findings translate into the uncomfortable character of Soviet urban life.

In the main text, there is little space for any emotions. I try to employ a rigorous approach for looking deeper into Soviet urban system. I rely very much on theories, models and statistical tools – perfected in urban geography during the decades of its 'quantitative revolution' – to penetrate inside the Soviet urban mechanism.

My preference is towards providing a number of dissections of that mechanism, each time with rigorously tested and quantified results. Such an approach is needed because the official Soviet statistical sources put at a disadvantage every student of Soviet realities. Many such sources are incomplete, vague and in some points unreliable. At a recent top political meeting in the Kremlin, G. Borovik, a delegate to the Nineteenth party conference, put the problem flatly: 'We are the people denied true statistical data . . .' (*Pravda*, 2 July 1988: 3).

My task was to uncover what is true by combining reliable bits of data, and tracing main trends, regularities, and tendencies, very much like consistency checks. Frequently I construct proxy variables to measure important factors of urbanization for which the Soviet statistical sources do not give direct measures. Proxy variables are possible, of course, only because other data or observations exist, and one may employ inferential statistical reasoning.

Big leaps of urbanization in the Soviet period

A broader outline of Soviet societal patterns will show how the topic selected for this book fits into the recent history of that nation. A discussion with elements of historical reasoning may serve this purpose.

Between 1917 and 1989 the urban population in the Soviet Union increased from 18 per cent to 66 per cent. Such rates are more impressive if one remembers the absolute size of the population involved: the number of Soviet urban residents has increased to 189 million in 1989 from the initial level of 28.5 million in 1917. An additional thousand of cities came into being to accommodate the avalanche of new urban dwellers.

As to the old thousand cities incorporated before the revolution of 1917, some made a big leap in size. Pre-1917 Russia had only two cities with population of one million or more – Moscow and St Petersburg; in 1989 their family is twenty-three, but none of these was created after 1917.

Around 4,000 places represent a category of urban centres of a type little known in old Russia. Each place is highly specialized and limited in size. It may be a growth pole of a local calibre, but each time investment decisions of the State are at its origin. They function as company towns, in essence, and contribute to a kind of suburbanization trend peculiar to the Soviet Union. A narrow range of job opportunities and an artificial, import-like, way of creating jobs make them incomplete cities, a fact implied in their generic name ('gorposelok' in colloquial Russian, or 'a settlement of urban type' in Bureaucratese).

Recently a number of nations in the Third World have manifested very high rates of urbanization. Since the Second World War, urbanization in Mexico, for example, has been comparable to the records set in the Soviet Union (Figure 1.1). Thus, the Soviet urban growth rates are no longer unique, their pioneering records were set within human memory.

Without degrading this topic to ideological debates, we may now ask, how original is Soviet urbanization? How special, individual and perhaps inapplicable elsewhere is it? A reasonable approach, that keeps the analysis in the mainstream of objective discussion, is to concentrate on urban development processes found only in the Soviet Union (or earlier, in the Russian Empire).

A number of Western scholars have collected evidence on the uniqueness or individuality of the socialist city, first started in the Soviet Union. Studies by Smith (1979), Demko (1984), French (1979), and Giese (1979), for example, attribute its uniqueness to the very mechanism that in 1988 is under fire in Moscow: centralized planning by the State.

Many but not all Soviet authors insist that Soviet urbanization is

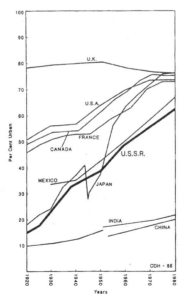

Figure I.1 Percentage of the urban population in the Soviet Union and other nations (1920–80). Notice the similarity of the curves for the Soviet Union and Mexico.

Source: Harris, 1988: 141

5

not only original but a model for the other nations: they followed the official line of ideology. They argue that the Soviets were the first to develop radically innovative cities that exemplify success for other nations. These arguments are hardly convincing now, in the middle of the crisis of the Soviet society. A recent study published in Moscow, provides a testimony about the absence of successes:

Many cities have declining population [and] . . . a number of economic indicators suggest that inequality among regions is growing.

Incorporation of new cities is declining as an evidence of stability in the network; . . . during the last decade this trend is clearly manifest.

Concentration of privileged jobs and lifestyles in a few bigger cities is hypertrophic and socially discriminating for territories within the nation. (*Razmescheniye*, 1986: 73, 74, 81)

With this evidence in view one may pay more attention to other Soviet authors: those who insist on their urbanization being unique without claiming its model-setting character.

This moderate position is voiced, for example, by V. V. Pokshishevskiy, a long-time doyen of Soviet population and urban geographers:

Cities of our nation may quite legitimately be considered as a specific type, as a world macro-region, very distinct from the others. (Pokshishevskiy, 1978: 142)

Indeed, there were powerful factors, spontaneously working long before the era of centralized planning in Russia, that are strikingly specific to its urban development. One may argue that Soviet centralized planning was no more than a forced adoption of these long-term spontaneous trends by the Soviet state.

A short schematic outline of pre-1917 urbanization in Russia may be in order here, because older roots of the process are important: this outline will serve as a counter-balance for the bias in main chapters, where only the current phase of Soviet urbanization comes to light.

Older roots for specificity of urbanization in Russia

Russia has a history with centuries of isolation from Western Europe, the cradle of modern urbanization The isolation was deep during the Renaissance and the period of great discoveries of the

New World. Muscovy's inland location gave little access to the world market and the early benefits of maritime trade. Participation arrived pretty late with the reign of Peter the Great (1689–1725), and continued up to the First World War. The post-1917 period saw a number of returns to the isolation syndrome, particularly during the decades of Stalin's rule. A diluted form of isolation happened as recently as 1980–5.

Obviously, the periods of cultural or political isolation brought opportunities for practising urbanization of 'a specific type' and in a form 'very distinct from the others,' as V. V. Pokshishevskiy said.

The Soviet tendency to be original in its ways and forms for urbanization also has another definite inclination: all must comply with the rules of austerity. This austerity originates in the size of the Russian Empire and the wastefulness of its political organization.

Since the time of Peter the Great, the Russian Empire spanned the continent, but it is essentially an inland empire, the biggest in modern history. Much of its climate is very inhospitable: only one square mile in ten is suitable for agricultural crops (compared to two in ten in the USA). The growing season is short in most regions, and where it is a bit longer, there is usually a lack of moisture also.

The outlay of resources for keeping the empire intact can not have been small, because its territory equals to one-sixth of all land on the Earth. Unlike other empires of modern times – Turkish, British, French, Spanish, Portuguese or Dutch – very few opportunities existed in Russia for employing the benefits of year-round waterways for its internal communications. Transportation costs for internal trade and imperial duty travel were a major reason the nation remained poor.

An additional and gradually growing burden for the nation was in the military style adopted by its rulers for keeping together the loose assortment of fertile lands and deserts, after they were 'consolidated' (also by military means) with Muscovy. For a long time internal trade could not develop to the point of cementing the nation, so it was logical instead to place garrisons everywhere.

Thus, historically Russia has found itself in the vicious circle of having overstretched resources and little room for adopting a more productive urbanization. Garrisons are at the origin of its urban network: they are the best connected nodal points, first by horseback mail, and later by telegraph and rail lines.

Interconnections are vital for unifying a nation that falls into eleven time zones. But in building unity, Russia adopted the most wasteful method, that of garrisons. Most cities of old Russia were originally military outposts, and until the 1917 revolution the main public buildings in urban places functioned very much like barracks: their uniformed staff had ranks conferred by the Czar. These included physicians, judges, teachers, university professors, surveyors, firemen, postmen, railway and other engineers. Merchants, before some of them expanded to became tycoons of industry or railways were licensed mainly as suppliers of the Czar army.

Little of the land was settled and well-developed, but the entire mass of the empire demanded the costly presence of the garrisons. Russia's neighbours in the pre-1917 period were also powers with military traditions: Turkey, Germany, and Japan. At certain periods Russian internal resources were so overextended and so thinly stretched over the empire that they were defeated in wars, usually due to failing in operations far from the heartland of Russia, as in the Crimean War (1853–6) and the war with Japan (1904–5).

This unique form of Russian urbanization was visible throughout the nineteenth century. The cities described in novels of that 'Golden Age' of Russian literature had pitiful architecture. The State decided how many public buildings each place could have, according to its administrative rank. Buildings had identical appearance, prescribed from the capital; all were very barrack-like whether they were hospitals, post offices or universities. Austerity was at the origin of orders given to architects. This austerity also affected construction during the railway boom and other capitalist ventures. Many of the ventures were at the mercy of State subsidies and decrees.

This type of urban development was very different from contemporaneous development in Western and Central Europe or in the Americas. In most Western urban nations there was no such dominance of the State. Cities originated and grew by the work of forces of commerce, with leading contributions from local initiatives.

Russia as the nation did not adopt the idea of *laissez-faire*. The Magdeburg Law (*Madgeburger Recht*), a cornerstone of European urban life, has no place in Russian cities. Few freedoms were permitted in garrison-centred communities. Russian traditions in politics suppress local initiatives, and approval from the imperial capital dominates. When private initiatives to build Russian textile industries started, they had more success outside the network of incorporated urban places. For example, Ivanovo, known as the 'Russian Manchester', was a village in the imperial roster of settlements in 1871.

To submit to the power of the local talents and wills present in each place is a foreign concept in Russia even today. Telling evidence about this point comes from *Moscow News* (July 1988, London edition: 9). In Kashin, a town of the Central economic region, a truck driver was threatened with arrest and licence confiscation for putting the slogan 'All power to the Soviets' on his vehicle. He did it as a protest against the red tape the local municipality faced in dealing with the pressing needs of the community. Instead of allowing local initiatives, a wasteful mechanism of state bureaucracy monopolizes decisions about all innovations. With such practices, expenditures for upkeep of the empire can only get larger.

All these traditions gradually translate into scarcity of material resources for developing urban centres. One particular trend has been endemic for centuries, up to the present time: the majority of Russian urban centres cannot afford the construction standards of the more fortunate nations, particularly those in the West. A very characteristic solution was adopted when newly-founded St Petersburg was developed as a new capital: the emperor simply forbade the use of brick or stone for buildings in all other cities, and that decree was in force for decades.

The austerity adopted for Russian urban development finds an interesting reflection in the works of S. M. Solovieff (1820–79), the famous Russian historian. He maintains that nature made its impact on Russian cities because most of them were located in well-forested areas, far from sources of building stones. Cities made of wood – and easily destroyed by fire – are characteristic of Russia, but in Western or Central Europe cities are built primarily of brick and stone. My study of Soviet cities shows that Solovieff's message, in one part of it, continues to be correct: the cities have indeed strong constraints in their life. However, the constraints of recent importance are, according to my observations, not in the poverty of building materials. Rather they are in the poverty of alternatives admitted on Russian soil into development of cities.

Chapter one

Hierarchy of cities: how it changes and why

The concept of hierarchy

When Geography takes part in Urban Studies it operates with messages coded as maps. It creates maps of deceptively plain appearance. They simplify land patterns of real life to keep facts at a minimum. Their purpose is like that of hieroglyphic symbols. Both emphasize the ideas communicated by images. Both try to bring concepts in a distilled form, without specific details. Which is justified, because one encounters enough headaches with mastering ideas behind images.

Mapped networks of cities are a case in point. Geography treats them with the concept of hierarchy. It is a tool to uncover reasons for the networks being what they are. A hierarchy, as dictionaries define it, is an arrangement of things (or persons) in a graded series; and it has lots of intricacy.

Examine any reference map of a nation, and make sure that it does portray a hierarchy of urban settlements. The map stresses various sizes and centralities for cities, which is a message about their grading in importance (Figure 1.1).

Maps show the scattering of cities of different rank and one may notice some regularity in it. Around any given city its nearest neighbours are mostly small in size. Big cities tend to be more widely spaced than small settlements. It happens because small places are numerous. The explanations may go deeper, and it leads to the treatment of urban hierarchies in Central Place Theories (Dacey, 1965).

There are two Central Place Theories, both started in Germany, before the Second World War. The origin was in a book by Walter Christaller (1933, 1956) and in another one, by August Loesch (1940, 1954). Both deal with emphatically distinct steps of ranking for cities.

Walter Christaller predicts a geometric progression in sizes of

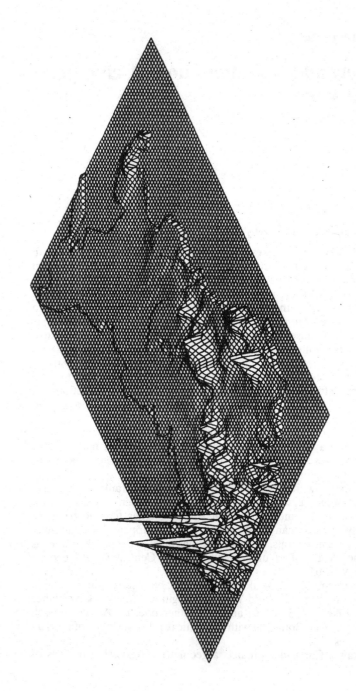

Figure 1.1 Soviet urban population, 1987, located in 250 main cities.

urban centres (Christaller, 1962). August Loesch admits also other levels of hierarchy. The disagreement is minor, if one compares it with significance of a common starting principle in two theories. Both Christaller and Loesch accept the making of urban hierarchies only by service functions of cities: by those addressed to rural areas and to vassal settlements. If a city succeeds in clustering such functions it grows in size. The clustering, in its turn, depends on the clientèle mass of serviced territories.

There are alternative explanations for urban hierarchies, with roots not just in urban services (Berry, 1961, Medvedkov, Y., 1966; Tinbergen, 1968; Vapnarskiy, 1969; Matlin, 1974). One of the explanations, the earliest of this kind, belongs to Auerbach (1913); it was rediscovered a generation later and very eloquently popularized by George K. Zipf (1941, 1949). It has potentialities which I am going to use in this chapter for clarifying features of the Soviet urban networks.

Zipf's approach to urban hierarchies is associated with a so called 'Rank-Size Rule'. It predicts city sizes in a nation without rigid and distinct steps in grading. The reasoning suggested by Zipf is free from postulating any unique privileges for services in cities.

If one sets aside numerous (and ingenious) attempts to portray the Rule as a clone, a derivative from Central Place Theories (Beckmann, 1958) or from trends observed for population density gradients (Okabe, 1979, 1987) the post-Zipf theory in Geography for the Rank-Size Rule is regretfully thin. It is unfair to the originality of Zipf's ideas, and there is more in the Rule than only an empirical and descriptive device for compressing data arrays.

In this chapter a view is presented that independence of Zipf's tradition is a virtue. I am introducing a train of reasoning for strengthening the Rank-Size model in its sovereignty. One may find appealing its freedom from rigid assumptions. It permits more room for experiments in analysing why real-life data fit so well the Rank-Size model, the fact well testified by many authors (Alperovich, 1984; Assamy, 1986; Boventer, 1973; Cassetti *et al.*, 1971; Medvedkov, Y., 1964; *Models*, 1970; Rashevsky, 1943, 1951; Richardson, 1978; Rosen and Resnik, 1980). Such features are just right for the job in mind.

Consequently, there will be opportunities to return later, within this chapter, to the Rank-Size Rule.

Why knowledge of urban hierarchy is important

The material so far discussed, in this chapter, presents evidence that Geography pays much attention to urban hierarchy. One may legitimately ask, if a justification exists for bringing the same approach into this book.

Why invest more into the same? What urgency is in learning the Soviet reality under the angle of urban hierarchy? Why may it benefit readers and give them a better understanding of social, cultural, economic and political problems in the Soviet Union?

One answer to this is in indicating that knowledge comes with structuring data into information. A process of learning has steps to replace vague opinions by solid findings. For that purpose there are specialized tools of research. They rely on concepts (like that of hierarchy). It means acceptance of traditions in studies; one benefits from their potential.

There is a certain urgency now in bringing the Soviet urban hierarchy into the limelight. Detailed arguments for that are in the next two sections. Here it may be sufficient to mention the present period of disarray in internal Soviet life. The nation is at a point of unpleasant discoveries about itself, and developments are worth watching closely. In any case, this nation is a nuclear superpower and an unexpected course of events within the Soviet Union may bring instability elsewhere.

One may specifically question the usefulness of the hierarchy concept. Why become preoccupied with it? What may it promise of inherent interest? Why does it fit the type of concerns about developments in the Soviet Union?

The answer to these questions is in expounding the meaning of hierarchy. The entire chapter deals with it. At the introductory level it is best, perhaps, to have help from relevant 'key words'. They are instruments of information retrieval. They are like co-ordinates for navigating among library shelves or in database browsing with a personal computer. It is easy to make a trial. It will uncover key words with close association to hierarchy.

There are pepper-and-salt relations for hierarchy. It goes together with systems, with flowcharts of authority in organizations. One examines hierarchy for learning who is the boss of whom. Other key words link hierarchy to functioning of systems. They also lead to ways of handling a system, to keeping it under control. This sounds like a journey deep inside the nature of the Soviets, into their rigid, military-like structures.

In examining the hierarchy of cities one may get closer to such key notions as 'societal system', 'national unity', 'poles of political power'. Let us remember that one of earlier books on urban hierarchy has the title *National Unity and Disunity* (Zipf, 1941). It is an acute angle for viewing the Soviet Union, where ethnic groups of Transcaucasia have been involved since the spring of 1988 in a noisy dispute over the territory of Nagorny Karabakh. While this event escalates to the level of a major constitutional crisis one may observe how it triggers many other peripheral regions to voice a spirit of dissatisfaction with the imperial order imposed by Moscow.

The regional structure of the Soviet Union has also economic contradictions and imbalances which may endanger the imperial dominance of Moscow. Central Asia dominates regarding the number of additions to the labour force, whereas Siberia has a monopoly on the main mineral resources. With their resources being clearly complementary the two macro-regions will gain most from establishing direct co-operation. But Moscow is more interested in the underdeveloped Central Asia to keep it in obedience.

One way of getting the answer is in examining the hierarchy of cities. The hierarchy develops from long-term favours in channelling national investments. Each urban centre functions as an accumulation of fixed assets, and in this sense an inquiry into the hierarchy permits one to assess the inertia of the existing economy, i.e. the subject of actual importance for the Soviet Union in its present phase of attempted, but slow going, changes.

Hierarchy of cities belongs to key Soviet structures. The Soviets are an urban empire. The Marxist ideology advocates urban growth. Ambitions to have industrial and military strength have been invested into cities. Much of the success or failure in these ambitions depends on the health of proportions among cities, how well they function in concert. It is acutely intriguing right now, because rapid growth of the Soviet economy is a matter of the past. During the 1980s it came to a grinding halt.

Now, the Soviets are debating reforms and blaming the former leader, Leonid Brezhnev, for short-sighted investments, and there are voices about 'paralysis of transport'. Disease of that sort is more complex than shortages of food. 'Transport product cannot be replaced by another and cannot be bought for currency' write Vasiliy Selynin and Grigoriy Khanin in Moscow, *Novyy Mir* (No 2, Feb 1987: 186).

There are other non-replaceable structures in society, like urban systems with a substantial distinction. However, urban systems are harder to modernize, it takes generations to build them.

Economist Abel G. Aganbegyan, a top advisor for the Kremlin, tells in a Soviet weekly how much more damaging is poor quality of manufactured goods. He admits that 1,500 factories are being penalized by the State: their goods of bad quality are not counted towards production goals (AP, July 21; Aganbegyan 1988). It represents a manifest decrease in industrial functions of cities. But what about tracing it in deeper roots?

This book has a chapter on economic functions of Soviet cities. It displays what is wrong at that level. Equally or even more, it is important to have the same for the roots, for the irreplaceable structure of the national urban network. It requires a process of visualization at the root level. The concept of urban hierarchy is a key participant in it.

Models and real-life experiences

This chapter (and others) combine two different types of reasoning. The first is a style with formal constructs of science, including mental models, like that of hierarchy or other mathematical models used for statistical data analysis. The second type of reasoning consists of informal and explanatory arguments, the more difficult of the two, because explanatory reasoning must interpret formal models, and derive conclusions from them or declare them invalid.

One enjoys objectivity and exactness in conducting experiments with formal numerical models. The results come as if automatically, and they are the same with any qualified operator of a computer. Broader life experiences of the analyst are practically irrelevant to the quality of formalized results. Explanatory reasoning is different. The background of an expert manifests itself very strongly in the quality of interpretations, which depend on who does the interpreting, what direct knowledge of the realities is involved, and how intimate was prior access to key structures. Here interpretations must compensate for weaknesses of the formal models. Such models bring results that remain rather indirect probes into major Soviet riddles and mysteries, and the output remains a method of remote sensing. These models need help from explanatory–interpretive reasoning, and conclusions are indirect.

The two styles of reasoning are familiar in many intellectual professions. For example, daily events in a modern medical clinic implement both types of reasoning. Patients are examined by laboratory tests, and a patient may be given laboratory results. This is all very exact, but it is not yet a diagnosis. More work must be done by the physician with the power of judgement, intuition and skill for inferential conclusions. Last but not least, the physician needs experience in either the same or similar cases.

In the interests of combining the benefits of both ways of reasoning, paragraphs in one style are interlaced with those in another. This format will, it is hoped, help avoid mistakes imbedded in shortcomings either of models or of informal reasoning.

Let us illustrate this by an example of how formal properties of models get in contact with real-life knowledge. Remember that Zipf's model pays little attention to service functions of cities, but such functions are basic in central place theories. Direct experience with Soviet realities permits one to see which of the two approaches corresponds to the facts.

The service sector in the Soviet urban economy is like a saying about the Swiss navy: It is a contradiction in terminology. Urban service functions are orphans when it comes to getting State investments. At the same time, non-State investors and operators are severely suppressed (Grossman, 1979). Punishment awaits persons who attempt to fill in the vacuum of non-existent urban functions. Soviet mass media call them 'speculators'. Soviet schools, TV and the press brainwash people to hate 'speculators'. They are scapegoats to be accused of keeping Soviet citizens in shortages or outright absence of consumer goods. Soviet courts deal with 'speculators' as with enemies of the State. Years of prison and confiscation of property are routine sentences for this group. Thus, total disarray in urban services results. Whatever services the State itself keeps are also under the constant suspicion of police, prosecutors and courts, because services require direct person-to-person relations; and in a system of totalitarian control such personal relations are threatening and alien.

Under Mr Gorbachev there is a new tendency towards private enterprise, on a small scale, aimed at upgrading the standard of living. People take the risk but feel very insecure, remembering the New Economic Policy (NEP) lessons of sixty years ago.

Because a description of urban functions in Soviet cities can hardly rely on service functions, chances for central place theories to explain Soviet urban hierarchy are minimal. In coming to this conclusion, we have a practical example of bridging abstract models together with realities of Soviet life.

The task of updating our knowledge of the Soviet urban hierarchy

Let us look at what kind of urban hierarchy exists in the Soviet Union and how it is different from nations outside communist rule. Are there many differences? What about upward (or downward) trends for features unique in the Soviet urban hierarchy? Have recent decades seen any radical change in the grading of cities?

Questions of this type come immediately to mind as soon as one faces the data of Table 1.1. One learns that Soviet urban centres have a pyramid-like distribution. Very few of the places fall into

Table 1.1 Soviet urban settlements in different size categories,
1926 to 1986

Population size (in 000's)	1926	1939	1959	1970	1986
under 5	1068	997	1747	2246	
5 – 10	378	761	1296	1430	3739*
10 – 20	253	501	798	919	1137
20 – 50	135	316	474	600	710
50 – 100	60	98	156	189	255
100 – 500	28	78	123	188	237
500 – 999	1	9	22	23	31
1,000 and more	2	2	3	9	22

Source: Gradostroitelstvo, 1975; Narkhoz, 1987; Sbornik, 1988.
*This figure combines the two categories of the smaller cities into a single figure.

the biggest-size class. Each step towards smaller size shows more
settlements.

Data like those in Table 1.1 are quite common in reference texts
dealing with urban geography. But they do not yet provide a
response to our questions raised earlier: they are raw material, not
well-structured information.

Clearly, the table exaggerates step-like grading of urban places.
It quite arbitrarily accepts one particular set of size classes for
grouping of the cities. It is hard to justify why such a set has priority over others. It is also questionable that the same size classes
have constant meaning over many decades. Because in 1926 a city
of 50,000 was, for example, closer in scale to the main city,
Moscow, than it stands now. For this reason, Table 1.1 has a biased
and incomplete description of Soviet urban hierarchies.

Unfortunately, answers for our questions are also absent in
quite advanced, classical studies. Knowledge in this field is now
less perfect than in 1970, when Chauncy D. Harris based his in-
depth analysis of Soviet urban hierarchies on the Soviet Population
Census of 1959 (Harris, 1970). One cannot safely apply older
results to current Soviet realities.

Almost a quarter of century separates us presently from the
latest cross-sectional Soviet data, which Harris analysed. Since that
time active transformation of Soviet cities has taken place, and this
transformation gives rise to a variety of speculations.

For example, the number of cities with populations of one
million or more has jumped more than seven times since the Soviet
Population Census of 1959. Table 1.1 shows a most noticeable and
dramatic change. Moscow doubled its total stock of dwellings
during the tenure of Leonid Brezhnev (1964–82). Roughly, this is

like creating another largest city, next to the old one. Similar progress has occurred in some other Soviet urban centres. New industrial centres came into being, for example, the two giants of the automobile industry on the Volga river. A bonanza of oil and gas fields in Western Siberia brought the Soviets into the world markets with a role comparable to that of OPEC. Soviet citizens with means achieved access to the private car.

Such developments have a potential to reshape urban life rather deeply. They might also transform the urban hierarchy. Chronologically, all of this has happened since the latest data base was analysed by Harris (1970). Re-examination of the Soviet urban hierarchy is very useful for catching up with these changes.

The updating task is not helped very much by the works of Soviet geographers, due to a differently placed emphasis in their research, which follows the stated goals of Soviet central planning. All Soviet research projects are under stern and vigilant political censorship of the Communist party. There are specific provisions for such censorship in the Soviet State Constitution. Thus, it is taboo to initiate a critical appraisal of planning efforts.

Intentions to change the Soviet urban hierarchy

Let us specify briefly, what after 1970 was of interest for Soviet geographers in the service of centralized planning. What did they want to change in the Soviet urban network? Is there a plan for a more equal grading of cities? Is there a desire to enhance the dominance of larger centres?

There are different schools of thought on this point in Soviet geography. The range of opinions is well displayed in two recent books, both of them collective monographs with contributions from representative authors. One is endorsed by the influential Moscow branch of the Soviet Geographic Society (*Problemy*, 1985). The other is 'Location of Soviet Population' (*Razmescheniye*, 1986). It came from the Moscow State University (MGU) Centre of Population Studies, and it has a wider circulation provided by the State Publishing House *Mysl*.

The first book advocates 'the process of centralization for all potentially centralizable functions' (p. 66). The other one insists on sticking to 'a tendency of forming a more even and uniform pattern of settlement' (p. 207).

It is probably better to place these two Soviet volumes into the category of political journalism rather than scholarly studies. The authors operate with incomplete data: to get a better picture one should look into a recent table on Soviet urban growth (Shabad,

19

1985, 1987) in *Soviet Geography: Review and Translation.*

What one may glean from the two books is mostly an orientation of interests and concerns. Their orientation is to promote modification of urban hierarchies; their intention is not to uncover patterns of logic in the actual processes of real life.

Basically, these two positions exhibit some conflict between them. One viewpoint is allied with planning practices on the cities with specialization on a national scale. Priority benefits must be for urban functions under the direct supervision of the centralized planning agency, under daily management of central ministries in Moscow. Interest in keeping the Soviet empire 'united and indivisible' are fuelling this sort of work. The reasoning of the authors, without saying so, serve the purpose of strengthening the dominance of Moscow bureaucrats all over the nation.

The second viewpoint is allied with local interest groups. It places the emphasis on bringing more urban places to the level of maturity. It is an attempt to foster a decentralized development, to create numerous poles of growth. Concern about the fate of smaller towns and projects to upgrade centres of intra-oblast districts are in this category of efforts. These are voices (with hesitancy and reservations, usually) who apply the logic of central place theories in reshaping Soviet urban networks.

Not all Soviet urbanologists fall into this polarization of viewpoints. There is much switching of loyalties to either of the two positions among Soviet geographers who are preoccupied with urban agglomerations. Some of them keep their orientation pretty fuzzy in this respect.

In much-advertised official projects for the Nationally Unified Settlement System (NUSS) there have been attempts to incorporate and reconcile proposals from both schools. It hardly helps to implement the NUSS. But the NUSS, in general, is with little implementation. The project is too ambitious, beyond the means of funding, at least as the Soviet State stands now, with the present decline of the Soviet economy. A review on the first ten years (Listengurt and Portyanskiy, 1983) of the NUSS cannot hide the fact that the NUSS is shelved; the text resembles an obituary. Now, when Gorbachev's administration has a negative stance towards the years of Leonid Brezhnev, it is unlikely to expect a galvanization of the NUSS. Meanwhile, real life brings other issues to Soviet cities.

Contemporary urban life spills very obviously over the rim of municipal boundaries. In the West many people who work in cities prefer residences in the suburbs or further out, in rural areas. The invisible arm of supply and demand opens up opportunities for that, bringing new functions into former rural settlements around major urban centres, both in Anglo-America and in Western Europe. Constellations of 'bedroom-places' are interlaced with networks of satellites that have complementary functions: jobs and services. In each case smaller settlements share the benefits of mature urban life: residents benefit, directly or indirectly, from job variety and from other attractions in the core city.

When similar development is advocated for Soviet suburbs, an observer of Soviet realities (and more so, a participant) is faced with a hard dilemma: to accept or to ignore the hardships of life in the suburbs. Soviet urban experts, as a rule, do not live in suburbs. It shields them from many worries. If they did live in the suburbs, it would be too difficult for them to focus their attention just on benefits and costs for 'the society' or 'the economy' on a national scale. Residents of the suburbs face quite another arithmetic on benefits and costs in their life, and the balance is very rarely positive.

Soviet urban agglomeration around Moscow, Leningrad and Kiev are quite chaotic in structure. Many suburbanites spend up to two hours each working day in job-related travel. Public buses serve residents within municipal boundaries, but they do not provide comfort and economy of time for millions of suburbanites. Their life is excessively taxed by abnormally long commuting. During weekdays it is hard to be on time at the job, and weekends are a rush from one queue in a store to another in order to get to sources of food. Such sources and services are available nowhere but just in the core city (with some exceptions in the Baltic Republics).

Life in Soviet suburbs does not benefit from the opportunities of owning a private car. Only a few families can afford them. The State sticks to the policy of keeping the private car in short supply and highly priced, and the support network of roads, service and petrol stations are very inadequate.

Lifestyles in Soviet suburbs are characterized by fatigue, constantly strained time budgets, and rudimentary services. The same may be true for people living inside municipal boundaries, but those in suburbs make a big jump up the scale of difficulties. As a result, the suburban population see themselves as outcasts; mostly they dream about getting apartments in the core city. It is exceptional to find families who left Moscow, Kiev or Leningrad for the suburbs by their own will.

Emerging urban agglomerations in the Soviet Union are the subject of study for many urban geographers in that nation. However, mainstream publications (Davidovich, 1976; Lappo,

21

1978; Listengurt and Portyanskiy, 1983; Lola, 1983) have emphasized new tasks for planning rather than addressing the actual
disarray of the realities.

In sum, one may discover different concerns and views in
planning intentions dealing with Soviet urban networks. A variety
of views could emerge between the 1960s and 1980s more easily
than before, because mass housing programmes at that period gave
more jobs for well-trained experts (those who had their own ideas
and a position for voicing views). Nevertheless, we have had only
scant assessment of what happens to the urban hierarchy.

The silence of Soviet experts may be very expressive. It indicates, frequently, a compliance with orders not to touch upon State
secrets. Anything may get into this mysterious category, Soviet
'state secrets', which provides a sure way to bag and keep in the
closet inadequacies of the communist system. Soviet experts are
absolutely silent when it comes to comparative assessment of the
Soviet urban network and those without centralized planning. For
answers, let us turn first to existing background knowledge, the
subject for the next section, and a step towards the findings of this
study.

Specificity of Soviet urban hierarchies

The hierarchy changed for all major Soviet cities over the period of
the late 1960s, 1970s, and early 1980s, dynamic and relatively
untroubled years for the Soviet economy. If Soviet planning can
shape the hierarchy, now would be the most propitious time to
leave an imprint on the grading of cities.

Let us try to discover the effects of planning in the grading of
urban networks, while leaving room for scepticism about such
effects. Soviet realities, as well as those of the Warsaw Block satellites, boast about being guided by the mechanism of centralized
planning, which happened to be not very successful. Let us see
how centralized planning works (if it does) in the area of hierarchies of cities.

The decade of the 1970s bore witness to a proliferation of
writings in Soviet geography on settlement restructuring. No effort
was spared by Soviet geographers, whose terms of employment
explicitly require them to contribute research results to planning.

However, it may be quite inappropriate to assume that such big
efforts necessarily sired results. Soviet experience frequently
demonstrates waste of such efforts. The pattern repeats itself again
and again, with each new political leader and in many fields. In the
end one starts wondering: is it not inherent for the Soviet efforts to

be wasteful? Or does the society have an ability to plan itself?

The Soviet system, generally, is notorious for forgetting about its major promises. It pledged, for example 'to overtake' the USA in per capita welfare. The main ideological document of the Communist party, its programme, once had a dream to provide, by the year 1980 for free food distribution, adequate dwellings for every family, free urban transportation, and so on. Instead, by then the Soviets arrived at a spectacular decline of their economy.

Average human life expectancy declined since the middle of the 1970s. All males in the Soviet population by the middle of the 1980s had a prospect of a shorter life: on average it is two years less in 1984–5 than it was previously (*Sbornik*, 1988: 118). Infant mortality in the Soviet Union has increased between 1970 and 1980 from 24.7 per 1,000 births to 27.3. In rural areas it was 32.7 in 1981 (*Sbornik*, 1988: 132), a level no better than in many Third World nations. Prices on commodities are rising. A tankful of petrol now costs a consumer four times more than in the 1960s. The very availability of daily food is now a matter of concern in practically every Soviet family.

Maybe the field of urban planning is a lucky exception with success: Soviet planning may influence the urban network. After all, only the State finances and supervises all construction projects of any size in Soviet cities. Only State officials are permitted to design houses, to construct urban blocks and whole cities. And there is evidence of sophistication in many Soviet writings on the subject of rational solutions for urban networks.

In respect to such expectations it is better to have solid facts, with answers based on objective analysis of mass data. Let us see if the Soviets have any manifestations of unique urban hierarchies. Is there a presence of specificity in the graded series of Soviet cities? What about the effect of glasnost, after decades of dictatorial rigidity in centralized planning?

The approach here will be to make use of international comparisons, to examine Soviet urban hierarchies against the background of others of a similar kind.

Comparative dimension: Eastern Europe in parallel

The findings will be more soundly based not just on Soviet data, but also, on data from Eastern European satellites of the Soviet Union. The statistical models employed rely on numerous observations of real-life phenomena. The larger the data set, the more reliable will be inferential conclusions derived from it.

One approach for assembling more observations is to take data from various years in the same urban network. This approach would not bring enough data for the simple reason, that one can not go too far into the past. Current Soviet practices have changed realities. Also, one needs to space out the years of observations, leaving some room for real-live events, which do not happen quickly. National urban networks do not change overnight. A reasonable period must pass between observations, for example, five years.

Thus, another approach becomes a necessity: collecting data from basically similar social systems. The best possibility for that exists in the Warsaw Block Eastern European satellites of the Soviet Union. Urban networks there also have experienced decades of command-type planning mechanisms. Each Warsaw Block country in Eastern Europe may yield a number of observations, each one related to a different five-year plan. By combining both approaches (with due regard to national distinctions in urban networks), one can obtain a database of desirable size.

Theoretical dimension: influences superior to planning

Comparative analysis of numerous urban networks may bring not just an updated or deeper understanding of Soviet realities. Next to it, and with quite its own significance, is a theoretical dimension – the study design, techniques of data analysis, and model building.

Such intellectual efforts are applicable, in principle, to any territory. Geographers do it in order to turn raw data into structured information, generally casting light on region-related or spatial patterns.

In a study focusing on several nations one has a chance to look deeper into hierarchy-related theory. Making use of these chances helps to make the findings more meaningful. First, it may provide a broader explanation for Soviet hierarchies. One may attempt to explain the grading of urban places, that Soviets live with, in terms superior and more lasting than their planning. It will be a search for an alternative origin of urban hierarchy.

Regimes come and go, but the land with its regularities remains and lasts. The study design employed in this chapter is aimed at making a further step into the spatial regularities of central place theories. Whereas many earlier studies looked mostly at the direction of urban functions, this one boils down to an explanation of urban hierarchies by characteristics of the territories that separate urban centres from each other.

The study is guided by an intuitive understanding that a certain similarity exists between a network of cities and a system of ponds interconnected by tubes. Consider the case when the tubes are different and small in diameter: some of them are silted, but all tubes also participate in draining the territory that separates the ponds. The territory is large; it belongs to different watersheds (similar to different economic regions within the Soviet Union). If some grading of volumes of water in ponds is observed, one can find a hierarchy of the studied objects, originating in the conditions of the tubes, in gradients of flows, and in the amount of water seepage from the territory surface into the tubes.

It can be a model for urban hierarchy which leads to data structuring with underlying explanations for a grading of places within an urban network. Maybe this model will strengthen independence for the Rank-Size Rule?

Input data: variables suggested by the Rank-Size Rule

In implementing this study of urban hierarchies, we first issued a warning about keeping away from distortions rooted in the preliminary grouping of cities, depicted in Table 1.1.

This table gave an impression that Soviet cities fall into clear-cut size categories, but these categories are an arbitrary artefact of a decision about the table's size! To be on the safe side, let us accept initial data in the most detailed form, individual city size.

At this point it must also be clear that the database departs from traditions of central place theories. The data will not refer to service functions of cities. One cannot, at any rate, ignore the simple and strong limitation that reliable portraits of service functions in Soviet cities are absent. What is left then? If central place theories fail us, we can try their competitor, Zipf's 'Rank-Size Rule'.

The first variable introduced into the analysis is $H(j)$ − number of inhabitants in each city, a traditional measure for the size of an urban place, well known by geographers. Sources for these data are available, and geographers know when to provide regular updates of $H(j)$. Also, the relatively high accuracy of such data is widely acknowledged.

One knows, certainly, that statistical services (Soviet and others) cannot be absolutely precise about the counting of residents in each urban place. Keeping this fact in mind, we round numerical values in the array of $H(j)$ for Soviet cities up to the nearest thousand residents. Finer scale rounding, up to the nearest hundred residents, appears in our arrays of $H(j)$ for those nations that serve for purposes of comparison. The procedure of rounding serves as a shield against initial small inaccuracies in data.

The second variable is j – the rank number of each city in its national roster, when grading of all urban centres is expressed in decreasing order of size. A subtle but important point to understand how deep is the difference in meanings of $H(j)$ and j. Putting the j-array on the base of the $H(j)$-array may leave the impression that both are the same, but they are not.

Theory of measurements, a discipline within mathematics, clarifies distinctions between $H(j)$ and j. The array of $H(j)$ is made of cardinal numerical values, and j-array consists of ordinal numerical values. Spacing of neighbours is a variable in the first case, but it is a constant in the other. Arithmetic operations which are correct for one array are inapplicable to the other. One cannot also return to $H(j)$ from knowledge of the j-array. There are also differences in meaning between the two variables. Notice that the meaning of $H(j)$ does not transfer into j. In volume, as well as in orientation, there is unequal content for $H(j)$ and j.

The first variable, $H(j)$, belongs to a category of self-standing parameters. It retains significance even in the form of an isolated numerical value, that of a particular city. It conveys the size meaning for urban life phenomena. For example, one gets from it the message that city X has 8 million residents.

The second variable, j, is meaningless when it stands alone. Unattached to a specific network of cities, it gives only the fuzziest knowledge. Consider an example: suppose that a career assignment in a transnational corporation brings you to the prospect of selection among several places, all in the category of 'the third largest cities' of unspecified countries. You would be really puzzled what to expect: the third largest city is a substantial centre in the USA or the Soviet Union, but not in a smaller nation, like Ivory Coast or Malawi.

Some common denominator appears only if one starts comparing the city with rank number three to other centres of the same nation. Then one may learn about the relative importance of such cities within each urban network. Important insight may follow if at this point, one discovers, for example, that in one network the third largest city is three times smaller than the largest city, whereas in another network it is twenty times smaller.

The variable (j) denotes a relative rank of importance; such measures require help from others of their kind within a network. Because it brings in itself the process of arranging objects in an order of grade or importance, (j) has direct relevance to the topic of urban hierarchies.

Additional data: territory characteristics

Additional variables are of interest in our intention to look deeper into the roots of the Rank-Size Rule. This study will link the urban hierarchy with characteristics of space outside municipal limits. The idea is to look at the organization of the national territory. Some indicators were selected to show how much cities enjoy favourable access to raw materials, food and energy sources. The grading of cities may be based on a somewhat parallel grading of opportunities to get materials for urban existence and growth.

The Soviet economy is very railroad-dependent. For this reason it is important to give attention to the density of rail lines. Soviet statistical releases permit one to see a trend in the densities of railway goods. If one knows how many tonnes of freight originate in a square kilometre of territory, this is very useful knowledge which permits describing the intensity of metabolic processes within the network of cities and their supporting rural spaces. An index of such intensity must by all means be included in the set of additional data.

Quite obviously, some cities intercept supply from other, less privileged urban centres. Thus the density of the urban network itself is also of importance: on average, it is likely that the more such density increases, the heavier is competition among urban centres.

Finally, there is an index of shape and clustering for urban networks. An index of shape for national territories is also quite informative: all are needed as soon as one attempts to compare different countries. Such measures help uncover to what extent conditions of centrality are rare. They provide a way to estimate the chances for cities to have very unequal benefits of location.

Altogether, there are five variables in the additional set of data:

(a) RG – density of freight transported by railroads (t/km^2, per year);
(b) RN – density or railroads (km per 1,000 km^2);
(c) U – density for the network of cities with population greater than 20,000 (per 10,000 km^2);
(d) SH – index of shape for the national territory (a ratio of the diameters of two circles, the circumscribed and the inscribed, on a map with national boundaries); and
(e) NN – Nearest Neighbour index, to uncover clustering or uniformity in the arrangement of cities on a territory.

NN-index is computed by a formula, which may be presented in BASIC notation as

$$NN = A/(0.5*SQRT(s/n))$$

where A = an average spacing of cities in km, s = area in km^2, and n = number of cities in the area. The numerical value of NN has a range from 0 to 2.15. The case of NN = 0 might occur if all settlements merge into a compact agglomeration. When settlements are with perfectly equal spacing NN increases to the limit, where NN = 2.15. Quite a remarkable point on the scale of NN values is NN = 1. In this case settlements are randomly spaced, in accordance with the Poisson probability distribution (Medvedkov, 1976; Medvedkov, 1978; Thomas and Huggett, 1980: 225).

It must be admitted that our set of additional variables represents a compromise. Plans to get more have been curtailed by scant Soviet statistical records. The availability of data also has constraints of another sort. Remember that the aim is to compare Soviet conditions with those of Soviet satellites in Eastern Europe, a comparison that demands parallel availability of data from several nations. Thus, expanding the additional data set is not very feasible.

Variables in this set are based on measurements and calculations of all space within national boundaries. Thus these variables may be used either for international comparisons or for looking at conditions of a particular nation in different years. The intention of this study is to do both.

Finally, it is important to emphasize that the values of the initial variables clearly change over time. One might ask, for example, how they are in respect to U and NN? What makes them change when time marches on? Is it not more correct to consider them as constants, specific for a given nation? In that case only international comparisons would benefit from U and NN, as happens with SH (index of shape for a national territory).

Both U and NN were computed in a form which makes them very sensitive to time. For that reason, a threshold size, 20,000, has been assigned as a minimum requirement for places participating in values of U and NN. The idea of the threshold size comes from observing that smaller places rarely possess the amenities of urban life in socialist countries. They lack investments to afford amenities or to produce goods for neighbouring settlements.

Remember that such economies function under rigid commands from centralized planning agencies. Investments are a monopoly of the State. Centralized planning does not afford splitting attention too far. That is why investments go only to large industrial units, i.e., places of substantial size.

The Rank-Size Rule as a hierarchy model

To find concise descriptions for urban hierarchy, one may rely on two data arrays: j and $H(j)$. A way for linking them together, as Zipf suggested, boils down to the simplest power function:

$$H(j) = H(1) \, j^{-1} = H(1)/j \qquad (1.1)$$

where $H(j)$ – population of the jth city, j – its rank number in a list arranged according to decreasing sizes of cities ($j = 1, 2, 3, \ldots, n$). This is the Rank-and-Size rule in its classic form.

It is remarkable that a model so simple as equation (1.1) continues to invite intriguing interpretations, but in the last fifty years they have come from experts of many disciplines. Examples are plentiful even in the Soviet Union, despite its generally conservative attitude towards innovation (Arapov and Shrieder, 1977). Their diverse interpretation come mostly from outside the field of geography, in such areas as general system theory, structural linguistic analysis, and evolutionary biology.

Geographers in the West sometimes ignore equation (1.1). It has no place, for example, in *Geographic Perspectives on Urban Systems* (Berry and Horton, 1970), in which everything about hierarchies is devoted to central place theories. On the other hand, a very recent example of the application of equation (1.1) to arrive at far-reaching conclusions is collective monograph edited by A.H. Dawson (1987). In its concluding chapter, this monograph evaluates results of spatial planning in Eastern Europe with the help of nine graphs (p. 332), all based on equation (1.1).

Such varying attention may indicate that not all is clear in equation (1.1). If geographers employ equation (1.1) they work mostly with its descriptive possibilities. It serves also as a useful device to help become orientated in bulky records on the number of residents in each city of a nation. Rather than search each record, it is much simpler to be guided by equation (1.1): the second largest city has half the population of the biggest city, the third largest has one-third and so on.

Equation (1.1) is interpreted very differently when geographers want to discover why and how the Rank-Size Rule works in real life. Such attempts have been made in Soviet geography, on the level of explaining the logic of power functions (Khanin, 1976). Simulation experiments have also reproduced equation (1.1) for those who are more at home with computers than with calculus (Matlin, 1974).

The mathematical logic of equation (1.1) is very transparent: cities keep constant ratios in size only if they have equal per-

29

centage of growth in the number of inhabitants. For the proof of this fact, consider events in two cities. Suppose that the initial observation, at time $t = 0$, discloses that their sizes are H(1) and H(2). At the next observation, when $t = 1$, the sizes may generally be designated as $k(1)*H(1)$ and $k(2)*H(2)$. Percentages of growth are $k(1)$ and $k(2)$; they are unknown at this point. If $t = 0$ and $t = 1$ have the same ratio of sizes, then:

$$\frac{k(1) * H(1)}{k(2) * H(2)} = \frac{H(1)}{H(2)}$$

which reduces to $k(1)/k(2)=1$. The last expression is, of course, equivalent to $k(1) = k(2)$, a simple answer that does not explain much of deeper concern for geographers. How does it happen that cities are initially of different sizes? Why do cities in positions of inequality switch into habits of having equal percentages of growth?

Without waiting for mysteries to have answers, some geographers use the Rank-Size Rule as an instrument for diagnosing systems-like families of cities within regional or national boundaries (Trus, 1977). Chauncy D. Harris went particularly far in this direction. He uses the Rank-Size Rule for delineating regional systems of settlements in the Soviet Union, and he has more trust in that rule than in official Soviet boundaries of economic regions or territorial production complexes.

In Soviet geography, there has been a very loud rejection of the Rank-Size Rule, in the form suggested by Zipf. Pokshishevsky advised indignantly against a regularity which attempts to be equally applicable for 'capitalist' and 'socialist' urban networks. In his criticism, Pokshishevsky (1978) also emphasized the poverty of the geographical content in equation (1.1).

On this point, however, Y. Medvedkov provided a rapid defence for equation (1.1) by expanding it along the lines suggested by regression analysis: he added two parameters with undeniable ability to measure the specific characteristics of the examined urban networks (Yu. Medvedkov, 1964). In this way a version of equation (1.1) was accepted in a number of Soviet studies as a tool for comparative analysis of national or regional urban networks (Gudjabidze, 1974; Matlin 1974; Dzhaoshvily, 1978). In the modified form it was possible to 'legitimize' equation (1.1), despite the ideological objections of Pokshishevsky, because comparative geography at that time badly needed tools for more accurate and objective judgements. A victorious recognition of formula (1.1), over ideological rejection, is documented in the *Soviet Geographical Encyclopedia* (volume 5).

The episode, minus the vexation of Soviet ideology, has broad meaning. Abstract theories do not receive much welcome in geography unless they march in close company with applications. The minds of experts are busy with empirical work and field data; their predilections shape standards of evaluation. Theories have support if they suggest tools of data analysis. If not, the most elegant theory may be ignored.

According to the expanded version one deals with:

$$H_j = K^{-1} H_1 j^{-b} \qquad (1.2)$$

or in BASIC notation:

$$H(j) = (1/K)*H(1)*(1/j)\hat{\ }b$$

where H(j) and j are the same as in equation (1.1), i.e. population and rank of a city (H, j),

K = primacy coefficient to signal a characteristic of the biggest city, a measure of its being within or outside the tendency for all other cities;

$b =$ the parameter of hierarchy gradient, numerically the same as a slope of regressing log H on log j (the log-transformation in this case permits the expression of the gradient in a linear form).

Differences between equations (1.1) and (1.2) are significant. With the addition of K and b, we no longer have an abstract formula but a signature for an urban network. Names given to K and b show that equation (1.2) is a device for sorting features of geographical significance in the hierarchies. It's of use in the discipline which is 'down to earth' specific.

To apply regression techniques, one treats formula (1.2) as an equation, making logarithm transformation of both its sides:

$$\log(H(j)) = \log(H(1)/K)) - b*\log(j) \qquad (1.3)$$

A linear structure of the type $Y = a - b*X$ is obvious in formula (1.3). If one has observations in two arrays, $\log(H_j)$, $\log(j)$, it is easy to use them as input for simple linear regression analysis. The algorithm of least squares permits one to determine $a = \log(H_1/K)$ $= \log H_1 - \log K$, as well as b of the equation (1.3). After that, with known H_1, one easily finds K, by computing, first, $\log K = \log H_1 - a$.

The initial application of equation (1.2) used input data on urban networks in twelve nations. It was a stratified sample representing a range of situations in the world at the beginning of the 1960s (Yu. Medvedkov, 1964). The main results from that study may be repeated here:

	(b)	(K)
Min. (China, Brazil)	0.738	0.693
Max (Austria, Chile)	1.629	2.757
Average	1.029	1.202
Standard deviation	0.259	0.602
Coeff. of variation, %	25	30

It is remarkable how close are the average values of b and K (1.029 and 1.202) to the idealized rule suggested by Zipf. Notice that equation (1.2) turns into formula (1.1). Zipf's model, as soon as $b = 1$ and $K = 1$.

Understandably, there is individuality among cities and urban networks in each particular nation; for that reason one is faced with the coefficients of variation 25 per cent and 30 per cent for b and K, respectively, in the world-wide sample of situations.

Model (1.2) has been rediscovered and very widely applied in various national schools of geography, including those in Italy (Fano, 1969), the Netherlands (Tinbergen, 1968) and Argentina (Vapnarskiy, 1969).

Comparative profiles of urban hierarchies

This section deals with findings which originate from working data arrays $H(j)$ and j. In other words, we examine here the most numerous and robust data, in accordance with equation (1.2).

By applying equation (1.2), one automatically turns the Rank-and-Size Rule on the track of comparative studies. Very much on the subject of comparisons, we now discuss the Soviet urban hierarchy with reference to other urban networks, including the Ukraine and the six nations of Eastern Europe, that belong to the Warsaw Pact.

The Ukraine and the six nations of Eastern Europe court straightforward comparisons. Their urban networks are close in size, the more easily to avoid questionable apples-and-oranges arithmetic, and to move step by step in reasoning. Initially, we match similar sized networks; next, we will learn what features in the hierarchy change when all Soviet cities are invited into the comparison.

Appendix 1.1, to this chapter, displays most of the numerical results of the work with equation (1.2). Appendix 1.1 highlights proportions in size for almost 3,000 cities. The Soviet Union alone does not have that many, but extended field for observations comes as the first clear advantage of adopting the comparative approach.

Each national network of cities has a number of observations in Appendix 1.1. Spacing between the observations is, in most cases, close to five years. Such spacing permits the registration of changes which might occur in urban networks because of purposeful strategies in centralized planning. It was also taken into account that planning agencies in the nations under study here operate mostly with the time framework of Five-Year Plans.

From Appendix 1.1 it follows that profiles of urban hierarchy experienced considerable change over the last twenty to thirty years. Each row in the table corresponds to an empirical formula, according to equation (1.2). Parameters (K) and (b) of the model are not at all constant in time. Altogether they disclose thirty-six situations with the hierarchy. One may observe a collection of experiences with urban hierarchy in a concise and standardized way.

Remember that the data items in Appendix 1.1 are fairly exact. Transformation of data into equation (1.2) brings additional suppression of inaccuracies, because the model specifies a general tendency for urban places, rather than individual conditions of each urban place. If some cities are underestimated in size and others are overestimated, and if both kinds of errors are infrequent, the tendency is not very much distorted.

Each of these urban networks has a family of rank-size curves (Figures 1.2–1.6). The curves reflect tendencies of growth for all cities with populations above 20,000, for the Soviet Union, where growth curves are given only for cities above 100,000 population. This exception derives from lessons learned from experiments with equation(1.2).

The equation does not permit seeing much below 100,000 if one looks at the aggregated urban network of a nation as big as the Soviet Union, because one must enter the range of values where logarithms have too little spacing. One works with graphs which plot logarithms of city sizes (in 1,000s) on logarithms of rank numbers, in accordance with equation (1.2). At some point on the graph, marks for individual cities are no longer separate, so not much may be learned about ordering of smaller urban places.

Some rows in Appendix 1.1 have numerical values K > 1. They signal that the largest city is overinflated; that is its size is

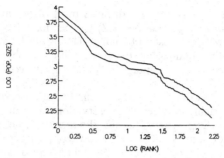

Figure 1.2 Soviet urban hierarchies according to the Rank-Size graph, 1970 and 1987.

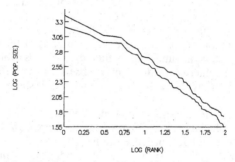

Figure 1.3 Ukrainian urban hierarchies according to the Rank-Size graph, 1970 and 1987.

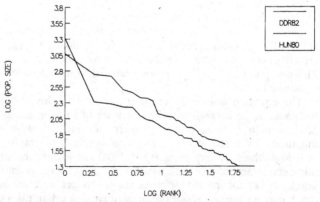

Figure 1.4 Urban hierarchies in East Germany, 1982 (upper curve) and Hungary, 1980.

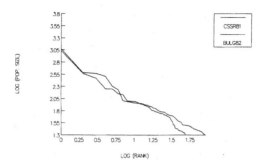

Figure 1.5 Urban hierarchies in Czechoslovakia, 1981 (upper curve) and Bulgaria, 1982.

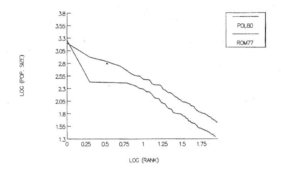

Figure 1.6 Urban hierarchies in Poland, 1980 (upper curve) and Romania, 1977.

35

above requirements suggested by the grading among the rest of the centres in the urban network. On the other hand, K < 1 means that the largest city is undersized for its supporting urban network.

Cases with b > 1 mean that the slope of the rank-size curve is very steep. To be exact, it is steeper than a hypotenuse in a right-angle triangle with 1:1 ratio of the side to the base. Less strong contrasts in urban sizes prevail, as Appendix 1.1 shows.

Additional guidance comes from three other columns of Appendix 1.1: (n), (R), (S).

The least exact information is in column (S), which shows how many steps in the urban hierarchy constitutes a deviation from a linear tendency uncovered with equation (1.2).

By consulting column (n) one learns how many urban places correspond to the discovered tendency for the studied networks. The bigger (n) in line of Appendix 1.1, the more reliable are the conclusions about K and *b* in the corresponding rank-size curve.

Finally, values in column (R) measure degrees of success in unveiling the tendency, |R| = 1,000 being the ideal success. As Appendix 1.1 shows, the correlation coefficient (R) increases all the time. This steady increase means that networks of cities have more and more internal organization, the one which finds its reflection in equation (1.2). Now consult column (R) at the bottom of Appendix 1.1, and notice how extraordinarily small (1 per cent) is the coefficient of variation for all computed values of R. This extraordinary stability suggests that one may safely rely on equation (1.2) for analysing proportions and changes in urban settlement.

It is striking how similar and high is the accuracy of equation (1.2) in all rows of Appendix 1.1. Notice, for example, that all absolute values in column (R) are very close to 1,000. It means that equation (1.2) approximates real-life data very well. It means also that (K) and (*b*) values do a good job of extracting tendencies from the initial data arrays.

The networks compared here belong to nations with the same type of command economy. The observed differences in (K) and (*b*) parameters of urban networks, shown in Appendix 1.1, hardly result from variations in economic or political foundations in those societies. The foundations are pretty much the same: there are similar over-centralized bureaucracies in metropolises, and they have similar intentions of suppressing initiatives in underlying strata of urban settlements.

All urban networks represented in Appendix 1.1 develop in conditions when the largest cities also have dictatorial political functions. Urban size and functions exhibit mutual adjustments –

one generates another, and there are feedbacks in mechanisms of growth. For this reason, K > 1 or K < 1 may indicate specific feedbacks, as well as resilience of other cities *vis-à-vis* the capital dictatorship.

Appendix 1.1 shows that parameter K varies much more than *b*. Its coefficient of variation, however, is lower than that for the sample of all nations of the world. The latter is close to 30 per cent whereas in these 'pure-socialist' conditions, the variation is almost doubled at 59 per cent. Coefficient K approaches its maximum in Hungary. It indicates that Budapest keeps its enormous size, despite all stated programmes dictating a more balanced development for other cities.

The minimum value for K is seen in the Ukraine, and the second lowest value of K is Poland. In both cases K < 1, indicating that Kiev and Warsaw are not at all prominent as political centres for underlying urban networks. Apparently they have other strongly competing centres: and, yes, in reality this is the case.

In its first years, Soviet rule in the Ukraine denied Kiev its former, centuries-long, function as a cultural and political capital. At that time the Soviets were safer placing their Ukrainian headquarters in Kharkov. Since then a certain priority in the size of manufacturing and in university-level education has remained in Kharkov rather than in Kiev.

Another factor is also detrimental for the political leadership of Kiev. In reality Kiev is very much under Moscow, serving as a cultural and economic centre, not for the whole territory of the Ukraine, but mostly for its south-western economic region. It shares influence with Lvov and Odessa. The population size of Kiev (in 1987: 2,544,000) is comparable to that of Kharkov (1,587,000), and not that far from the size of Dnepropetrovsk (1,182,000) or Odessa (1,141,000).

The cultural influence of Warsaw in Poland is substantially challenged by Krakow. There are also strong regional centres with competing influence in manufacturing: cities like Lodz, Wroclow, Posnan and Katowice. The history of the Polish labour movement in the 1980s proves that Gdansk also may compete with Warsaw in political influence.

In the national urban system of the Soviet Union it is striking to observe a change for Moscow leadership: notice how K > 1 is replaced by K < 1. This may mean that the initial political monopoly of Moscow in the Soviet Union is being more and more challenged by regional capitals. There may be other, rather numerous, explanations for the fact that the urban network is switching into K < 1.

Soviet Urbanization

The Soviets obviously have difficulties running a city as large as Moscow: 8,967,000 residents on 12 January 1989. Together with its satellite towns there are 14 million urban residents in the Moscow metropolitan area. All this mass of people live far from sea coasts, in an area with poor natural resources.

Moscow is a railway dependent metropolis, like Paris or Madrid. However, it differs from Paris and Madrid in having delivery distances of prohibitive length. Daily bread, energy and raw materials travel a thousand miles before they get to Moscow. Agricultural lands around Moscow have poor soils, and the climate is harsh. There are no important mineral resources nearby.

To make all things worse, the 1980s are a crisis period for Soviet railways, because they are in a run-down state. During the 1980s they have systematically failed to comply with the targets set in the National Plan. Railway support is vital for Moscow, but it has deteriorated because of decades of neglect. Every winter brings railways to a point of collapse; supply for Moscow is becoming more and more of a problem for Soviet authorities.

Further constraints on Moscow's growth originate in Soviet lifestyles. Labour is so cheaply paid that all adults in every family must be breadwinners. Shortage of housing plus rigid rules for employment and establishing residence make it impossible for families to live near jobs. As a result, Moscow residents are heavily taxed by long commuting distances.

The physical layout of the city and the suburbs creates many chaotic features. The spatial structure of urban functions does not minimize commuting to jobs or for shopping. Perennial shortages of consumer goods make it necessary for local residents to allocate at least as much time for shopping as their jobs require. In general, life in Moscow is getting more and more stressful.

The Soviet Statistical Office hides exact data on a sudden upsurge of mortality in Moscow during the 1980s, but the phenomenon is common knowledge in the circles of Soviet intellectuals.

All these problems together decrease Moscow's ability to provide the type of leadership which in 1959 was signalled by an indicator $K > 1$. Gradually this city has slipped into a diminished position of cultural and economic leadership, and it now dominates only the European part of Russian Federated Republic (RSFSR).

. In Czechoslovakia, Romania and Bulgaria, capitals seem to have more political and economic monopoly than Moscow may presently succeed. This conclusion comes from K values present in Appendix 1.1. The three nations are maintaining K at a level which is about 1.5 times higher than in the USSR.

Like the Soviet Union, East Germany shows declining leadership in its political capital relative to the national urban network. Appendix 1.1 reveals quite a spectacular declining trend for the K-parameter of East Germany. This trend conforms with practically all known processes in the economic geography of East Germany.

Recent decades have seen a restructuring of the urban hierarchy in East Germany. From a case with $K > 1.3$ there developed a case with $K < 1.3$, meaning that East Berlin is losing its political leadership. As an economic centre of East Germany it cannot surpass Leipzig, nor can it function convincingly as a cultural centre of East Germany. Artificial isolation between East and West Berlin – the infamous Berlin Wall – ruins the official party line about making East Berlin a viable and strong political capital.

When the K-parameter is getting smaller it means that the leading city (East Berlin) is well behind its subordinated settlements in growth rates.

More light on compared hierarchies comes as soon as one starts examining column (b) in Appendix 1.1. Generally speaking, one can hardly expect dramatic variations for numerical values in array (b). The purpose imposed on the b-parameter in equation (1.2) makes it more likely to exhibit a phlegmatic behaviour.

Technically, numerical values of (b) originate from fitting a smooth curve to data points in a two-dimensional co-ordinate space. The points reflect individual features of cities, but nothing individual penetrates into (b). It is just a gradient measure for a sloping linear curve. Thus, column (b) in Appendix 1.1 is unlike column (K), with its filtered-out individualities of cities. For this reason, one cannot expect a range of values for (b) those similar for (K).

The furthest spaced numerical values for (b) in Appendix 1.1 belong to the nationwide urban system of the Soviet Union ($b =$ min) and to the Ukraine ($b =$ max). This great difference may be explained, quite obviously, by features that are also vastly dissimilar for the two territories: one includes all Soviet land and the other the Ukrainian part of it. This reasoning leads to the conclusion that gradient parameter (b) of equation (1.2) may be sensitive to size of territory and to the level of its development. At this stage no more than a hypothetical statement, this conclusion must be examined in a rigorous way. The following sections do that.

The very unco-ordinated behaviour of the b-parameter among all the Warsaw Pact nations is intriguing. They are all centrally planned. Copying of planning goals from the Soviet Union is quite common. But one who watches the jumps of the numerical values in column (b) is little reminded of central planning.

Only East Germany and the huge network of cities of the Soviet Union shows similarly monotonous upward trends for the b values. But these two nations exhibit the widest dissimilarity in almost all characteristics of cities. They are widely apart in sizes of national territory, in accumulated investments per unit of land, in railway or urban network densities.

The behaviour of the Ukraine's b value does not follow the pattern of the overall Soviet urban network, but is more like Czechoslovakia's which b jumps up and later goes down. In Poland and Romania the opposite succession of the jumps occurs: first down and then up. In the hierarchies of Hungary and Bulgaria in which the behaviour of b goes up-down-up.

Considering all the trends in (b), the patterns do not suggest operating a purposeful planning of urban hierarchies. There is too much fluctuation within nations, without any apparent correspondence with changes of political leaders. For example, most of the b-jumping occurs in Bulgaria and Hungary, where changes in leadership were few or none in the years covered by Appendix 1.1.

The mysterious behaviour of (b) calls for an explanation. The subsequent analysis attempts to provide it.

How territory characteristics influence urban hierarchy

Two sets of initial data have been considered. The first of the sets has yielded results (Appendix 1.1). They have mysteries, as previous section reveals. The second data set, presented in Appendix 1.2, may clarify the behaviour of the urban hierarchy, specifically, the strange jumps in the b-parameter values.

For this purpose, let us turn now to steps for analysing Appendices 1.1 and 1.2 in parallel. Arrays in the tables are of a different nature; we may call them 'Zipf' (Z) and 'Geography' (G). Appendix 1.1 has 'Zipf' indicators, and they portray specific attributes of urban hierarchies. Appendix 1.2 displays indicators of another sort. As the word 'Geography' implies, the G-set has parameters of space where cities function.

Both Z and G are multidimensional arrays, or matrices. To link one matrix to another we employ the Basic Canonical Correlation Model (Levin, 1986). It has its origin in the works of Hotelling. One of his contributions, 'Relation between two sets of variables' (1936), specifies exactly what we need for arrays 'Zipf' and 'Geography'.

One array may be related to another by an equation:

$$Z = F (G) \qquad (1.4)$$

Canonical Correlations suggest a way for determining F of equation (1.4), and for this purpose there is an algorithm to compute canonical indexes:

$$I(G) \rightarrow I(Z) \qquad (1.5)$$

The indices are remarkable: they have optimally selected weights, $w(i)$ and $W(i)$, which operate, in this case, in formulae:

$$I(Z) = w(1)*K(i) + w(2)*b(i) + w(3)*n(i) + w(4)*R(i)$$

$$I(G) = W(1)*NN(i) + W(2)*SH(i) + W(3)*U(i)$$
$$\qquad + W(4)*RN(i) + W(5)*RG(i). \qquad (1.6)$$

Notations for variables in equation (1.6) are the same as in Appendices 1.1 and 1.2.

In computations of canonical indices there is room for models with different levels of complexity. This means accepting some or all of the variables and urban networks from Appendices 1.1 and 1.2. Each case must lead to matrices of equal size, to fit the requirements of matrix multiplication. There is a preliminary stage of computing Pearsonian product moment correlations for columns within and among Appendices 1.1 and 1.2. Next, one deals only with square matrices which have the property of symmetry. These make matrix multiplication and inversion easier. Present availability of megabytes of computer memory in PCs makes this computation problem less acute. But in Moscow experiments with Appendices 1.1 and 1.2 created a headache having to bother constantly about committing to memory the limits in available, much dated, machines.

Experiments with equations (1.5) and (1.6) have a very straightforward logic. First, one deals only with similar-sized networks of cities, and next, one removes this condition. The former case permits only Ukrainian cities to represent Soviet conditions. The latter case permits all Soviet territory (with all cities from 100,000 up) to be present in $I(G)$ and $I(Z)$. Table 1.2 explains it in more detail: participating variables from sets G and Z also have variations.

The experiments disclose considerable influence of territory characteristics on urban hierarchy. Central planning or no planning, location patterns of urban centres and levels in urban and railway density are dominant in establishing the urban hierarchy. The main parameters of the hierarchy, (K) and (b), are illuminated as soon as one traces, with equation (1.6), how key influences are

41

formed and channelled. This evidence suggests that the dynamics of (*b*) and (K), the essentials in urban hierarchies, are in good correspondence with the 'Geography' set of variables.

Table 1.2 Experiments with canonical indexes

Test with $I(G)$ $I(Z)$	Levels of complexity		Components in indexes	
	In networks	In indices	$I(Z)$	$I(G)$
(1)	Low: similar sized networks, 24 cases	Low	K, b	RN, RG
(2)	High: 26 cases, including the USSR	Low	K, b	RN, RG
(3)	Low: similar sized networks, 24 cases	High	K, b n, R	RN, RG U, SH
(4)	High: 26 cases, including the USSR	High	K, b n, R	RN, RG U, SH

Table 1.3 shows the evidence. There are three successful models, each one allowing us to learn how much (K) and (*b*) from the 'Zipf' set of variables are dominated by territory characteristics, by the legacy of centuries of history. For that one compares coefficients in equations.

First, there are three important coefficients. They play a role in making the equations numerically correct, and at the same time they participate in measuring the level of success in modelling. Absolutely perfect success might occur if the principal coefficients go up to unity, and at the same time, the canonical solution, equation (1.4), is nearly perfectly one-dimensional: it does not require other versions of equation (1.4) with residuals. In this experiment each time residuals are small, and, considering them as negligible, one may associate all 'explained' variance with levels of the principal coefficients. This line of reasoning, is described by

M.S. Levine in terms of 'redundancy analysis for canonical correlations' (Levine, 1986: 23-6).

With this type of canonical solution, it is possible to accept the principal coefficients as proxies for functional equivalents to multiple correlation coefficients. In social studies it is a success if a multiple correlation for a model attains the level of 0.7. One is usually much constrained by noise in initial data from getting that high or further. Squared values of the principal coefficients indicate, roughly but logically, the percentage of total initial 'variance' which this model captures. Let us accept the advice of M.S. Levine to enclose the word 'variance' in quotation marks because, strictly speaking, variance for sets of variables (like 'Zipf' and 'Geography') is different from the traditional definition for a single variable.

Sidestepping mathematical subtlety, many analysts of societal data will accept the 0.7 level in a multiple correlation as a sign of capturing roughly half of all signals in the data. The square of 0.7 is equal to 0.49 or to almost 50 per cent in respect to total variance in data, as soon it is standardized to unity (the latter condition comes automatically with employment of inter-correlation matrices). To clarify half of real-life complexity is not a small thing at all; it is a sort of threshold success in exploratory studies.

Notice that the bottom section of Table 1.3 portrays quite a spectacular adequacy in modelling. It is on the level 0.9, i.e. close to the absolute success. What is left unexplained by the model is comparable with likely inaccuracies in initial measurements and counts. There seems little chance to attribute anything to direct actions of purposeful planning, which are all outside our model. They are not specified in Table 1.3. The bottom line here suggests little need to bother with fishing for influences of planning: such influences have negligible effect on the urban hierarchies considered here.

The variables in Table 1.3 have the same notations as in sets 'Zipf' and 'Geography'. They permit easy reference to the material in the Appendices 1.1 and 1.2. There is, however, one step in data processing which one must keep in mind: all numerical values enter Table 1.3 equations upon scaling by the usual statistical means. They are transformed into deviations from the average and then divided by the square root of variance; this transformation creates clear independence from the initial units of measurement, which may be very arbitrary in raw data.

The models displayed by Table 1.3 are quite explicit. They indicate precisely what influences and consequences are made of. Each side of the equations may harbour tight internal interplay of

Table 1.3 Success in explaining hierarchy by territory characteristics

Test type	Hierarchy parameters and their weight	Success measure	Territory characteristics and their weights
1	0.3982*K + 1.16654*b =	0.7218 *	(0.0322*RG + 0.9718*RN)
2	Test 2 yields a non-significant, low, canonical correlation. This version of modelling fails. With access only to RG and RN in the 'Geography' set, one cannot clarify adequately well (K) and (b) behaviour for all cases in Table 1A		
3	0.315*n − 0.853*K − 0.799*R + 0.582*b =	0.8468 *	(2.274*RG − 2.088*U − −0.975*SH − 0.516*RN)
4	0.0022*n − 0.973*k − 0.938*R + 0.835b =	0.9048 *	(1.400*RG − 1.698*U − 0.853*SH − 0.309*RN)

Note: Test types are specified in Table 1.2.

the specified factors and consequences, but this does not show the nature of the influences. Like it is always exactly urban hierarchy, behind K and *b* in the left hand side of the equations.

With the selected indicators for the 'Geography' set one may make separate trials to observe, first, the power of (RN) and (RG), which means an influence from railways. Next, (U) and (SH) were added, for arriving at conclusions about more of the influence coming from the location pattern of cities. It is interesting that whatever is communicated by (U) and (SH) can replace separate influences communicated by the (NN) indicator. Remember that (NN) is a tool to measure the presence of regularity, randomness or clustering in urban networks. Because on the level of employed aggregation it turned out to be without key importance, the column of (NN) from Appendix 1.2 could not make its way into successful models. Thus, (NN) does not have a place in the equations of Table 1.3.

As always happens with correlations, one must be careful about making cause-and-effect statements. Arguments for them are of the type of informal reasoning. Help comes from designing a time-lag for the arrays 'Zipf' and 'Geography'. Correlations alone might suggest that influences are directed not just from right to left in the equations of Table 1.3.

However, causes are usually placed in time somewhat earlier than the present. The future is not powerful in changing the past: it does not have the certainty of tangible legacies. There is no clear backward engineering for nations in making the present occur with images of future prospects. Political slogans may suggest that possibility, but the closely-lopped actors of real politics deal mostly in expediency. In matters like urban hierarchy, too much of the future depends on decisions of actors yet to come, on unknown major innovative capabilities in technologies or social structures. Who could predict with certainty that the Soviets in Russia in 1917, for example, would reschedule even a private life?

This reasoning has a place in the study design: many entries in Appendix 1.2 have a time lag with respect to Appendix 1.1. They permit attaching the meaning of causes to the right side in canonical equations. In Table 1.3 it is the side with variables from the 'Geography' set.

Why urban hierarchy is hard for planning

It is a well known fact that the Soviets and their Warsaw Pact allies have made many attempts to remodel urban networks (see also pp.22–4, in Soviet works, and elsewhere in the literature).

Western scholars and practitioners of physical planning view with respect and sometimes with fascination the multitude of ideas in their field, about which they read or hear from Soviet or Eastern European planners. Other results of this attention may be examined in two impressive volumes: *Planning in the Soviet Union* (Pallot and Shaw, 1981) and *Planning in Eastern Europe* (Dawson, 1987).

No doubt about it, the Soviets have a large number of busy planning institutes attached to 'Gosgrazhdanstroy', the State monopoly agency, that produces and implements blueprints for residential complexes of any size. Only a fraction of the projects go under the leadership of Moscow Central Institute of Urban Research and Physical Planning. And that consortium alone has, as its former director, N. Belousov, frequently explained it, twenty-eight or even thirty hard working institutes. There is more in Leningrad, and also in Kiev, Minsk, Vilnus and in all other capital cities of the Soviet Union Republics (Iodo, 1985).

In a decade spent in research in the Soviet Academy of Sciences, intimate knowledge of many planning endeavours was gained by the author. It was possible to register quite a respectable level of expertise during duty travel visits to the majority of planning institutes in the Soviet Union Republics. Among colleagues busy in applied work, it is frequently clear that the position of front-runner belongs to experts in urban planning. They lead both in the number of talents employed, and in their earnest desire to do something beneficial. The same impressions came from duty travel to Poland, or in joint studies with colleagues from East

Soviet Urbanization

Germany, Bulgaria, Hungary, and Czechoslovakia.

Yet it was also quite common to hear, in a private way, admissions from the hardest-working experts how discouraged they are with real-life results. These comments, perhaps, were the origin of the present attempt to trace the impact of planning on the urban hierarchy.

The findings communicated by equations in Table 1.3, make it plain and firm: planners are helpless, their job only a pretence of being in command of the urban network. Events go on, the main influences coming not at all from planning, but derived from internal forces in existing urban networks. These other forces are in command, together with influences from territorial organization by rail mainlines. The planners are like people in small boats when a hurricane's whirl dictates events. Currents and gales play with the boat while the unlucky sailor, disregarding that its propeller is missing and the rudder is lost, painfully attempts to start an engine.

Let us examine the findings to see how to improve the effectiveness of planning. What prohibits the equations in Table 1.3 from having some constructive meaning?

Some of the factors highlighted in the equations are, after all, quite accessible for purposeful changes. In principle, investments may change both railway density (RN) and urban network density (U). It is even easier to manipulate RG, because more tonnes of goods per unit of territory may be prescribed as planned targets. It involves less capital outlay than RN and U require, and it resembles many earlier prescriptions of planning: more tons of steel, more of coal, more electricity, more DDT on cotton fields and orchards.

The equations in Table 1.3 have the appearance of numerical tools employed for conclusions of the 'what if' type. Planners in finance matters and engineering do a great deal of work with such tools. They are standard procedure among personal computer users, giving prominence to programs that automate 'what if' reasoning. Why not do the same with the equations that link 'Zipf' to 'Geography'?

1. By beefing up RG for a jump of one standard deviation, and with U, SH, RN values held constant at their zero averages, a force comes to the right side of the equation; the force has magnitude $1.267 = (0.9048*1.400)$, which in the left side of the equation, induces a jump down for K a shift in the scale of K values, where all are marked in standard deviations of $1.302 = (1.267/0.973)$. We assume that n, R, b may stay constant.
2. If SH, RN, RG are held constant but U is forced up one

standard deviation, then *b* must step down, provided that *K*, *R*, *n* are held constant; in its travelling down the numerical value of *b* must cover 1.84 units equal to its standard deviation, because $1.840 = (0.9048*1.698/0.835)$.

3. A change in the urban network density (U) is 5.5 times more influential in manipulating the hierarchy (*b* or *K*), if one compares it with changes of railroad density (RN) and assumes equality of efforts for stepping up one standard deviation U or RN; it comes from comparing the canonical coefficients 1.698 and 0.309.

All statements, 1–3, sound like music of triumph for a planner's ear. They appear to be exactly the guidance requested from science. The problem is that everything is wrong with the reasoning in these statements. Interpretations like these are criticized in texts on canonical correlations: they are 'dangerous and misleading' (Levine, 1986: 18). It will be a disaster to act as 1–3 suggest, notwithstanding all arithmetic evidence above. The arithmetic is accurate but logically inappropriate.

The difficulty arises from the nature of real life. The more one has success in applying the canonical correlation model to a process, the further that reality stands from prospects of being accessible to planning based on arithmetic reasoning. Real life does not fit 'what if' thinking because of the inseparability of components, a property that simply prohibits predictable manipulations of the components. The problem is how to handle that.

Imagine taking a spoon and lifting it off the surface of a table. Suddenly you notice that the spoon is attached to the place, it tips the plate, pull others towards you, and in a moment all the soup is on your knees. It is a nasty trick if somebody plays it. Exactly similar and even more awkward things must happen with urban hierarchy and urban systems: the accomplishment of the canonical model discloses it.

Canonical correlations succeed only when they find tightly linked things. For such realities it is wrong to apply the rules of arithmetic because of dominant multicollinearity in variables, as statistical analysis puts it.

Table 1.3 shows a tight linking of factors on the right side and a similar tight linking of consequences on the left side. No procedure of measurement permits knowing for certain how influences branch among consequences, so separating influences is very difficult. One cannot, after all, keep things constant, as suggested in statements 1 to 3 above.

On the other hand, these troubles with the tightly-linked factors do not mean that nothing is clarified by Table 1.3. Signs of canonical coefficients are unlikely to be wrong, and absolute values of coefficients are, in this case, safely far from zero. Thus we can accept, for example, the following statement:

> In general, by increasing the rail network density (RN) it is possible to play down the primacy of the biggest city, as measured by (K), or to increase contrasts in the grading of other urban centres, as measured by (*b*) growth in rail goods (RG) can also produce the same outcomes.

This observation casts light on the mystery surrounding urban hierarchies, that was registered by Boventer (1973). He described the roughly equal effectiveness of the national economies in France and West Germany, despite clear contrasts in the structure of their urban hierarchies. It was an illogical finding, because one expects that the hierarchy of economic centres matters. For France and West Germany there was no sign of removing a gap in numerical values (K, *b*). Now we know that a sort of mutual compensation in the behaviour of (*b*) and (K) marches with economic development in nations. This mutual compensation benefits either the largest city or less prominent centres.

Other interpretations of interest are suggested by the principal coefficients in Table 1.3. In their case the multicollinearity of variables does little harm. On the contrary, the more numerical precision the better the values in the following observations:

1. Urban systems in six nations of Eastern Europe, when we view them in comparison with the Ukraine, demonstrate a range of features in the hierarchy (K, *b*), where roughly half of the variance (52 per cent) is explained by railway networking and usage (RN, RG). This conclusion is suggested by the numerical value of the principal cofficient, presented in Test (1) in Table 1.3 presents it.

2. If all Soviet cities participate in matching of hierarchies, explanation (1) is invalid. The unacceptably low value of the principal coefficient results from Test (2): $R(can) = 0.157$ comes as a response to a step down in railway densities, after inclusion of Soviet territory. Likewise, it is a consequence of big 'holes' in the Soviet urban network.

3. With an increased number of territory parameters, and in particular, upon measuring territory shape (SH) and density in urban networks (U), we may claim a more complete explanation for the urban hierarchy – up to 72 per cent of variance in arrays (K, *b*, R, *n*). They relate to comparisons of urban systems with similar size.

4. Finally, we have the hardest version of comparisons. To look at all Soviet territory adds difficulty. Other urban networks are much smaller, but surprisingly, the highest level of success comes in these equations. Notice how big is the principal coefficient: $R(\text{can}) = 0.9048$. Possible coverage of the variance in arrays may jump up to 80 per cent. The paradox of finding the greatest success in seemingly hardest comparisons can, however, be explained.

This paradox relates to the earlier-noted quite peculiar properties of the canonical model, which works best with tightly-linked variables. If one looks at it from another viewpoint of planning, it becomes clear that version (4), certainly, has the most complications for planning. The complexity and size of the entire Soviet territory make it hardest to impose planning approaches. Remember the fallacy of the 'what if' reasoning: at this point it is clear that the 'what if' fallacy exhibits a step-wise increase as soon as the Soviet territory and cities enter the analyses. The Soviets were the first to develop ambitions for an all-embracing, command-type planning. It appears, according to the findings, that they have the most hopeless conditions for such a job.

Presence of random walk in planning

There is a mathematical model, familiar in studies of random processes, which by tradition refers to the experiences of an intoxicated person who attempts to reach home from a lamp-post, with darkness around and in mind. There is no information which side is home. Trial-and-error tactics do not work because all is quickly forgotten. After a number of steps in one direction, the person hesitates, stops and makes another attempt at selecting a direction, unrelated to the previous one.

I am not going to connect that random walk model with experiences of Soviet planners in a formal way. Instead, as a conclusion, it will be useful to get a graphic image of developments with the hierarchies. Let us have a look at paths walked by urban systems, those in the hands of the Soviets and of their Eastern European allies.

The graphs must reproduce reality in the form which is suggested by the most successful model: from Test 4, according to Table 1.3. It permits squeezing most from data arrays 'Zipf' and 'Geography', and for all urban networks under consideration. Soviet conditions are in it twice: first, within the Ukraine, and next within all Soviet territory.

Technicalities of the canonical correlation model permit creating, in an optimal way, paired indices, such as equations (1.5) and (1.6) show. Table 1.3 provides each observation point in time with two indices, one for conditions of the territory and another for conditions in the urban hierarchy. The indices make use of familiar empirical material. A row from Appendix 1.1 has numerical values for portraying a particular urban system at some year; it transforms into $I(Z)_{tk}$, where t refers to the year of observation and k refers to the name of the territory. In a symmetrical way, values from a corresponding row of Appendix 1.2 permit determining a paired index, $I(G)_{tk}$, that describes the territory conditions. Both indices together produce a point in an appropriate co-ordinate space.

Figure 1.7 illustrates the result of plotting the paired indexes. It is a standard X-Y graph. Because it relates already-named variables of use in canonical analysis, let us call it a Z∗-G∗ canonical graph. Its ordinate and abscissa accept numerical values taken from sets 'Zipf' and 'Geography', but first initial values must be the statistically standardized. By using the star marked symbols, Z∗ and G∗, we are reminded of that specificity. The Z∗-G∗ canonical graph has a quite objective scale for both its co-ordinates. Standard deviations for computed values of the indices bring markers for the abscissa and ordinate. The paths that urban hierarchies project on the Z∗-G∗ co-ordinate space do not look at all like a purposeful and determined progress. Systems of cities which belong to the same territory are encased into outlines. Some are waving, like counter-lines around meandering rivers. Arrows next to the outlines show changes in the direction for path segments. Shifts in direction are frequent.

Generally, a movement along the Z∗-axis indicates deeper differences among various provincial cities, and a simultaneous decline in the leadership of the main city. Steps to the right along the G∗-axis mean facing more immaturity in territorial organization: cities are wider spaced, densities are less impressive for the rail network and for goods generated by a unit of territory. All is more like it was in the past of nations which now have their paths within the left, and privileged side, of the graph. Mnemonically, the co-ordinate space has a succession of conditions similar to those of the Soviet territory. The latter has a more mature organization mainly along its western fringe, and steps northward along that fringe bring you to Leningrad, a former capital with declining leadership.

The pattern for all collection of paths is predetermined by the canonical correlation model. The paths exist within a ribbon,

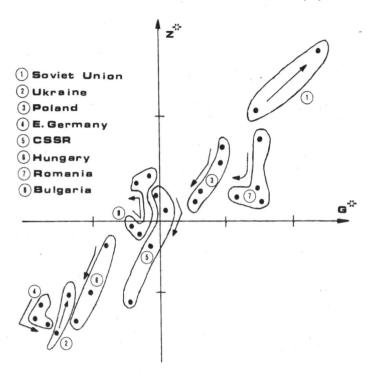

Figure 1.7 Reaction of urban hierarchies (Z∗) on changes in land utilization (G∗). The ticks on co-ordinate axes correspond to 0.5 of standard statistical deviations for the values of Z∗ and G∗. Arrows show the direction of changes. The values of Z∗ and G∗ are computed according to formula (4) in Table 1.3.

shaped like a very elongated 'cloud'. The ribbon is orientated along the 45 degree diagonal in respect to the axes Z∗ and G∗. Such structure emphasizes that Z∗ has very tight relation with G∗ ($R_{can} = 0.9048$).

The paths for Poland, Hungary and the Ukraine fit very strongly into the general ribbon orientation. Consequently, the observed behaviour of urban hierarchies is in the mainstream of uncovered tendencies. Most of our comments have closest applicability to all segments in the paths of Poland, Hungary and the Ukraine.

The paths of Czechoslovakia, Bulgaria, East Germany and Romania most resemble a random walk. They have sharp turns.

51

There were periods when evolution of their urban networks for a time departed from the usually packaged tendencies, traced within all assembled cases, in the material of Appendices 1.1 and 1.2. Invariably, there were later returns to the mainstream.

Directions walked by individual urban systems in the co-ordinate space of the Z_*-G_* graph always result in a vector; one can get an idea of the preliminary existing parallelograms of forces based on these vectors if one examines computer printouts showing the expanded structures of indices Z and G. These permit observing what territorial characteristics prevailed for turning particular segments of the path one way or another. It is possible now to describe the most frequently repeated situations:

1. There are segments in the paths indicating a deepening of the contrasts among provincial cities that is paralleled by restraining the growth in the main city; this process is accompanied by an increase in the density of rail goods (examples: the USSR, up to 1976; Ukraine and Bulgaria – from 1965 and during the 1970s).
2. More uniformity comes at other times to provincial centres, and as they lose individuality, the prominence of the capital of the nation grows; all studied factors of territorial organization contribute to this process (examples: Poland, Hungary and Czechoslovakia during the 1970s up to 1976).
3. The same evolution is observed for urban hierarchies but contributing factors are less equally strong, the most prominent stepped up tonnes of rail freight per unit of territory (East Germany, and Romania during the period from 1965 to 1976 and Bulgaria from 1960 to 1965).
4. Some segments in the paths which bring more diversity to various provincial centres, and a simultaneous decline in leadership of the capital city, where the growth rate slows; the major factor contributing to increasingly diverse territorial organization is the addition of more cities to the national network (Czechoslovakia from 1965 to 1970, Bulgaria, from 1970 to 1975).

Because of the tight linking of variables, we must be warned that real life is not as simple as the above interpretations imply. Further turns in the paths are difficult to foresee, because development does not disclose purpose. Urban networks with apparently the same system of planning develop remarkably opposite trends in their paths, for example, in the Ukraine and Poland.

The final and main conclusion may be stated in quite plain words, because it is definite: this is not planned development.

Appendix 1.1

Urban hierarchy profiles in the Soviet Union and in other Warsaw Pact nations: Analysis based on equation (1.2)

Territory	Year	Parameters of equation (1.2) primacy	hierarchy	number of cities	Rank/Size correlat. coeff.	Rank/ Size steps
	(t)	(K)	(b)	(n)	(R)	(S)
Soviet	1959	1.2515	0.7548	146	0.9868	3
Union	1970	0.9757	0.7627	221	0.9877	3
	1976	0.9116	0.7806	254	0.9880	4
	1979	1.1325	0.7808	273	0.9868	5–6
	1986	0.8530	0.7830	290	0.9854	5–6
Ukraine	1959	0.4685	0.9653	135	0.9945	7
	1970	0.4246	1.0033	175	0.9943	7
	1979	0.6490	0.8867	47	0.9888	7
	1985	0.6902	0.8714	48	0.9874	7
Bulgaria	1960	1.6308	0.8841	32	0.9865	6
	1965	1.5870	0.8858	37	0.9896	4
	1970	1.4369	0.9308	39	0.9917	5
	1974	1.3260	0.9110	43	0.9898	6
	1982	1.2673	0.9253	48	0.9862	7
Czechoslovakia	1960	1.6743	0.8972	51	0.9831	6
	1965	1.5911	0.9053	51	0.9872	6
	1970	1.6150	0.8792	59	0.9889	6
	1976	1.6177	0.8308	76	0.9884	6
	1981	1.6172	0.8000	88	0.9904	5–6
East Germany	1965	1.3089	0.8066	104	0.9880	7
	1970	1.2437	0.8089	107	0.9911	7
	1975	1.1767	0.8170	112	0.9948	6
	1982	1.2103	0.8185	116	0.9963	5–6
Hungary	1960	4.0454	0.8251	45	0.9436	6
	1970	3.6505	0.8440	54	0.9627	6

Urban hierarchy profiles in the Soviet Union and in other Warsaw Pact nations: Analysis based on equation (1.2) *Continued*

	1976	3.3829	0.8394	61	0.9706	6
	1980	3.1752	0.8464	63	0.9742	6
Poland	1960	0.6900	0.9356	114	0.9963	6
	1965	0.7134	0.9250	129	0.9970	7
	1970	0.7239	0.9054	148	0.9976	7
	1976	0.6909	0.8956	173	0.9975	7
	1981	0.6570	0.9025	187	0.9965	6(?)
Romania	1960	1.9046	0.8965	47	0.9657	4
	1965	1.8844	0.8680	57	0.9715	5
	1970	1.6815	0.8804	73	0.9785	5
	1977	1.4767	0.9003	94	0.9860	8(?)
Minimum	—	0.4246	0.7548	32	0.9436	3
Maximum	—	4.0454	1.0033	290	0.9976	8
Average	—	1.4538	0.8654	105	0.9858	—
Standard deviation	—	0.8666	0.0589	—	0.0113	—
Coeff. of variation	—	60	7	—	1	—

Appendix 1.2

Characteristics of territory organization: Soviet Union and Warsaw Pact nations of Eastern Europe

Territory	Year	Nearest neighbour distance	Shape of area	Urban Network	Densities Rail Network	Rail Goods
	(t)	(NN)	(SH)	(U)	(RN)	(RG)
Soviet Union	1959	1.04	7.07	0.3339	5.6	84
	1976	1.04	7.07	0.4656	6.2	161
Ukraine	1959	1.45	3.40	2.235	35.1	839
	1970	1.45	3.40	2.899	36.5	1309
Poland	1960	1.44	1.88	3.642	86.0	916
	1965	1.44	1.88	4.121	85.9	1090
	1970	1.44	1.88	4.728	85.3	1220
	1976	1.44	1.88	5.527	85.4	1473
East Germany	1965	1.32	2.73	9.630	147.0	2400
	1970	1.32	2.73	9.907	135.0	2440
	1975	1.32	2.73	10.370	132.0	2680
Czechoslovakia	1960	2.02	5.32	3.984	103.0	1645
	1965	2.02	5.32	3.984	104.0	1720
	1970	2.02	5.32	4.609	104.0	1850
	1976	2.02	5.32	5.937	103.0	2120
Hungary	1960	1.38	3.04	4.839	108.0	1025
	1970	1.38	3.04	5.806	99.0	1260
	1976	1.38	3.04	6.559	90.0	1420
Romania	1960	1.64	1.70	1.975	46.2	326
	1965	1.64	1.70	2.395	46.2	471
	1970	1.64	1.70	3.067	46.2	720
	1976	1.64	1.70	3.865	46.5	958
Bulgaria	1960	1.29	2.92	2.883	37.1	346
	1965	1.29	2.92	3.333	36.9	505
	1970	1.29	2.92	3.513	37.8	620
	1974	1.29	2.92	3.874	38.8	700

Note: Calculation of indices is explained in the main text. Periods with observations terminate in Appendix 1.2 earlier than in Appendix 1.1. It permits to incorporate time lags, into experiments with projecting data of Appendix 1.2 on those of Appendix 1.1.

Chapter two

Soviet cities and their functional typology

Urban typology and its applied meaning

Functions of cities, indicated by employment percentages, are sometimes an end point of research in economic geography (Harris, 1943; 1945; 1970; Nelson 1955; Alexanderson 1956; Crowley 1978). Urban employment data fit in theories on location of industries and illustrate the concept of urban or regional specialization. Such data also furnish a plethora of colourful signs for school and college maps that display the mix of industries in cities.

The subject of functional typology is an organic next step in data compression. It is easier for memory and logic if the range of employment proportions in cities fall into a small number of basic patterns. A few clear patterns can redirect attention to regularities, to essentials behind groups of similar places. In this chapter, the term 'functional typology of cities' refers to the uncovering of groups of features common to Soviet urban centres. A method of measuring similarities in employment percentages of individual cities is used to reveal this typology.

However, the typology is in no way our end point. It is, rather, a tool. Basic patterns in the functions of Soviet cities will shed light on two very sizeable national problems of the Soviet Union.

The first is the problem of overloaded Soviet transport, already discussed in Chapter 1 in reference to the example of Moscow. Obviously, the Soviets can no longer hide instances of paralysis on their railways, and what they admit may be only the tip of the iceberg. It is reasonable to expect that the lion's share of transport overloads come from the functioning of cities. The types of cities may explain deeper roots for Soviet transport problems, those of a structural and lasting character. To reveal structural problems, we can use the typology method to uncover frequencies of long-distance freight requirements for Soviet cities.

Second in order of discussion, but not in emphasis, comes the problem of the dated composition of the Soviet economy. That economy was too long busy with the phase of industrialization that was completed in the United States by 1900. Erroneous concentration on yesterday's basic industries created a number of unpleasant surprises. One of them surfaced as a sharp decline in sources of growth. By the middle of the 1970s the Soviets were obviously too much behind the time with their dominant technologies. Returns on efforts in industries dropped drastically, and funds were no longer available for investment (Hardt, 1987; Ofer 1987; Panel on the Soviet Union 1987; Aganbegyan 1988).

There was no room for parallel financing of butter, guns and development. In 1975, the Soviets made a milestone decision, they stopped funding development. 'The rate of growth of fixed investment falls dramatically from an annual rate of 7.0 percent in 1970–75 to an annual rate of 3.4 percent in 1975–80' (Levine, 1983: 161). Towards the 1980s they had no more butter. The Soviets cannot tolerate the same fate for guns.

Because of that, since the early 1980s, they have been giving the highest planning priority to post-industrial technologies of the Information Age. But are there signs of success? Does the emphasis on information technologies have an impact, for example, on the growth rate of urban centres?

Soviet cities have something in common with ancient Egypt: like the Pharaohs' pyramids, huge Soviet factories absorb much toil, but they bear little to their builders beyond serving as temples to the dogma of industrialization. The Soviets do not destroy their old factories in their current attempts to enter the information age. Consequently, it is quite logical to measure the magnitude of the inert force of the old urban functions. Do outdated industries dominate the economy and submerge innovations?

To answer that question, we will measure how much of the current growth pattern of Soviet urban centres is explained by the old urban functions. To do so, we will isolate a part of the variance in the growth rates of Soviet urban centres during 1970–86, that very part which is a response to the functional typology of cities of 1970. If the response is a dominant share of the variance then it means that the Soviets move like a prisoner chained to a cannon ball.

Sampled cities and their starting point profile

The task of arriving at a typology of Soviet cities kept the author quite busy for a number of years (O. Medvedkov, 1975; 1976; 1977). Those studies revealed that a rigorous approach to urban processes has better roots if a constant sample of cities is examined at various points in time. For this purpose we use a sample of 221 Soviet cities which had at least 100,000 residents by 1970.

At the end of the 1970s the Soviet Statistical Office released, for research staff of the Soviet Academy of Sciences, a limited number of print-outs from the 1970 Population Census. This release of data permitted access to unique mass data on employment percentages in all 221 Soviet urban centres. A 1970 cross-sectional portrait of functions in Soviet cities may appear to be irrelevant now, after eighteen years. But the old profile of cities is the only solid information of this kind: nothing similar emerged in 1979, after the next Census. The 1970 data can be employed to trace influences from the past on current Soviet realities.

Although the 1970 data are not current, it is possible to use them as a key for very useful findings. These 1970 data make up a 'legacy of industrialization profile' of Soviet urban centres. We will compare that legacy with current overloads of Soviet railways and with growth rates of Soviet cities. Both comparisons will help clarify the extent to which current realities carry an imprint from the starting point functional typology, the legacy of industrialization.

Specialization of cities and how it affects transport

Geographers traditionally look at cities as structural nodes in the territorial organization of society. Two important aspects of nodes are specialization and integration. The transition from specialization to integration is roughly equivalent to creating links between individual economic activities. The radius of the links depends on the proximity of the nodes and on the presence of any barriers in the space between them.

It is well-known that, over time, specialized cities change. From a highly specialized economic centre there may be development to a multifunction node. At the same time, many cities retain their specialization. Such cities remain internally unbalanced. Because they can function only in concert with others, they must overcome distance in a particular region, within the national territory as a whole, or into foreign countries to establish trade connections in order to meet their basic needs.

The Soviet economy, however, keeps inter-urban links within the closed space of national boundaries. This limitation is accompanied by adopting artificial prices (protectionism policy) and by giving priorities to military contingency plans. Consequently, the national frontier acts as an outer and absolute barrier for economic links. But the absolute nature of the Soviet national boundaries also simplifies our task in reasoning about the most probable distances which Soviet cities employ for their interaction.

Although long-distance freight may be justified in some cases, generally costs can be saved with short range co-operation among specialized cities. Theorists in Soviet economic geography insist on the economy of short-range links. They give a colourful display of the potential benefits of so-called Territorial Production Combines, socio-economic complexes and integrated development of regions in general (Saushkin, 1980: 237-58). But what are the real-life links that make urban co-operation possible? How are they distributed?

Answers to these questions are fragmentary. What is known suggests lack of measures to comply with the theory about short-distance links for specialized cities. Soviet cities have to rely on the limited facilities of existing railways, and they do not behave parsimoniously in respect of them. For example, according to data in the Soviet statistical yearbooks, 'Narkhoz' the mean length of haul on Soviet railroads was 861 km in 1970, 923 km in 1980, and by 1986 it had grown to 940 km (with a slightly bigger figure, 941 km, in 1985).

The increase in the mean length of haul, to which every tonne of Soviet rail freight contributes, was 35 per cent from 1940 to 1986. The railways account for 69 per cent of all transport work in the Soviet Union, if the total 5,551 billion tonnes/km (1986), disregards the pipelines because they distribute mainly raw material but not the products manufactured in cities. The other modes of transportation on service for cities also had a rapid increase in distances travelled by each tonne of goods. The inter-city truck hauls are not specified in the Soviet statistical yearbook. Yet it is possible to find some data in other sources. The absolute volume of inter-city truck hauls rose twelve-fold from 1940 to 1960, and by two-and-a-half times from 1960 to 1970 (Golz and Filina, 1977). The total volume of rail freight in 1986 was 6.7 times what it was in 1940 (Narkhoz, 1987: 343–4).

By an appropriate grouping of 'Narkhoz' data it is possible to show that the average railroad travel of goods in inter-city exchange grows more rapidly than it happens for goods originated in rural areas of the Soviet Union (Table 2.1).

The network of railroads, major channels to connect Soviet cities, was essentially stable all this time. More reliable figures on that are not in 'Narkhoz' (which tends to give inflated values, by counting all rail track outside mainlines), but rather in Soviet college texts on economic geography. According to N. N.

Table 2.1 Railway goods contributing to functioning of Soviet cities, 1960–86

Categories of goods	1960	1970	1980	1986
A Goods with origin in the intercity exchange (steel and pig iron, trucks and other machines, processed food,' other goods')				
billion t/km	390.1	746.5	1143.4	1290.7
million t	416.5	653.6	869.5	931.3
average km	937	1142	1315	1386
B Goods with origin in rural areas (grains, timber, ores, coal)				
billion t/km	712.5	999.9	1224.3	1341.2
million t	841.2	1144.4	1329.4	1424.6
average km	847	874	921	941
Ratio of A/B for average km	1.1	1.3	1.4	1.5

Source: Narkhoz (1987): 343–4.

Baransky (1956) the post-war mainlines of the Soviet rail network totalled 120,000 km, and the most recent text (Rom, 1986) shows a total of about 144,000 km. Those estimates suggest an increase of 20 per cent, or two times less than the growth of the Soviet population during the same period.

Between 1940 and 1986 the railways' share of Soviet freight dropped from 85 per cent to 47.5 per cent, but this decline reflects progress with modes of transport which do not link cities: many pipelines were built to export Soviet oil and gas, and truck freight, mostly short-range – within cities and suburbs – was stepped up.

The design of urban typology

Eight types of specialization are considered in this study: manufacturing, construction, railway and other transportation, trade, health care, education and culture, research and development (or 'science' as the Soviets define it), and last but not least, administration and economy management. The universe is all Soviet cities that had 100,000 or more residents by 1970. The conclusions about city specialization come from an indirect indicator: percentages of employment. Such data portray the inputs in specialization rather than the end result.

Of the three factors of production, labour, fixed capital, and raw materials, only the first participates in our typology. However, the other two factors usually have tight correlations with the first. In Soviet practice, these correlations have the strength of legal norms.

There is a rigid State monopoly for allocating labour with capital and raw materials. These allocations are made with simplistic and universally applied yardsticks. Soviet investment decision rules are standardized. Many of them are in the 'Standard methodology for determining the economic effectiveness of capital investments', and they have been in operation since 1969. Decrees of 1974 specified a uniform coefficient of effectiveness for all Soviet investors, which were invariably only State agencies (up to very modest exceptions, introduced by Mr Gorbachev in 1988). For the purposes of this study such rigid investment rules have the virtue of supplying a solid justification for the extended interpretation of employment percentages.

The consensus in urban studies seems to be that one may ignore the incompleteness of data on employment as indicators of cities' functions. It is a silent assumption that it would be a mistake to ignore an opportunity of dealing with available data, particularly when hardly any substitute data exist. But employment percentages have limitations, they invite more sophistication in the analysis. Specifically, it is desirable to amplify the features of the adjustment of cities to each other. Thus we will emphasize the measures that worked for the entire sample of cities.

In the current case, attention is paid to differences in the density within the network of nodes with particular specialization. We will look at three critical levels in the development of each specialized activity: local significance, regional links, extra-regional links.

This typology involves the following steps:

1. An array of employment data for each individual city is transformed into a vector of eight elements with scaling of the elements in a way that would make it possible to differentiate cities against the background of the entire range in the USSR.
2. Specialized sectors are identified for each city to give an idea what it produces in excess of its own needs and the needs of its surrounding region.
3. The links are characterized by which a city must make via railroads, with other centres that employ fewer people in a particular sector.
4. Inferential reasoning permits finding for each city a number of sectors that make it a participant in overloads for the railways,

this process takes into account the most probable radius of links for each type of specialization in the city, and considers the density of nodes endowed with the particular specialization.

5. Despite the technicalities in step (4), it is important to realize that our spatial analysis compensates the relatively narrow data on labour inputs: the radii of urban interaction reflect both the impact of labour and the degree of proximity in a network of cities.

6. Conflicting trends are highlighted in directions of specialization, by paying attention to balances or imbalances between secondary and tertiary fields of the economy. (The secondary field represents material production, given priority by Soviet planners of industrialization; the tertiary economy has less burden of heavy freight and easier acceptance of post-industrial innovations).

Probabilistic reasoning on distances in inter-urban connections

An explanation may be needed to clarify this approach to special-ization. The nationwide field of observation employed in our urban typology requires us to omit details normally included in studies of particular industrial nodes or territorial production complexes.

In this chapter the term 'specialization' is used in an unusual sense. Specialization is measured on the basis of employment in aggregated fields of activity: manufacturing, transport, trade and so forth. This approach is aimed at helping us to learn about the more lasting structures of stone and traditions in each city.

The aggregated fields of activity are also selected for the contrasts among their effects on the evolution and functioning of cities. The evolution element surfaces in commitments to continue with investments already undertaken, with specialized infra-structure, for example. An acute dimension of urban functioning is dependent on specialized sectors on uninterrupted long-distance freight flows.

Operationally, our urban functional profiles are shaped into numerical indices and eight-element vectors. The aggregation is patterned after approaches used in econometrics in the study of macroeconomic processes (Schatteles 1975; Johnston 1980).

Specialization is seen here as the ability of a city to transform a given amount of resources so as to meet the needs of the other places. Specialization thus implies complementarity. Specialization refers here to the geographical division of labour across the entire set of cities under study. A large percentage of employment in a particular sector of the economy is therefore of interest to us, and any contrasts in employment patterns correspond to the ability of nodes to generate links and zones of gravitation.

Integration is observed when a city contains several equally significant specializations. If the percentage of employment in the *i*th sector is equal or close to average for all cities in the nation, then one may logically assume that the *i*th sector are present in any particular city on the level of satisfying its needs. It is because the Soviet Union lived for a long time in a state of autarky, particularly in 1970, the date of the 'profile of industrialization legacy'.

This approach is obviously schematic. It ignores regional differences in the productivity of labour. However, broad schema are required if we are to find results that lead us to some solution for the problem. This study deals, after all, with large cities in which both the working habits of the population and the availability of modern equipment tend to be evened out.

The existence of a territorial complex assumes the presence of complementary levels of specialization in cities in close proximity to one another. In this case, the focus is on proximity. Proximity will depend on the density of distribution of a particular level of specialization. A sparse network of places results in much longer links, making the formation of a complex less likely.

Such an interpretation makes it possible to consider tying specialized activities together into a territorial complex. It does not pretend to accomplish much more. The schematic approach becomes evident in the probabilistic determination of the radius of links. The radius was determined on the basis of the density of the network of places endowed with a particular type of specialization.

There could, of course, be situations in which the subtle requirements of local industries create links, not with a nearby city that appears to offer a basis for complementarity, but to a place thousands of kilometres away, in another economic region. Such situations had to be ignored. In limiting the study to cities that had reached 100,000 in 1970, less populous satellite places that are integral elements of territorial production complexes had to be ignored. Our focus is on the ability of large cities to work in concert with other cities in the same large territorial complex.

The mechanism of specialization is quite complicated. Without all the aforementioned simplifications, it would be difficult to arrive at simple, clear-cut computations.

Let us now assume that developed territory consists only of the cities and their zones of gravitation. Immediately, it makes sense to distinguish two types of zones. A resource zone is a set of places that supplies a particular activity in a city; a zone of influence is a set of places that are supplied by the results of the labour in that city. Each specialized activity in a city gives rise to two such zones, which rarely coincide. Multinodal and interlaced zones predominate.

A particular urban place is likely to have links with several cities, but because different activities are represented in each city, no two cities are likely to generate zones comprising the same tributary places. A city with a commanding position in one type of specialization may be subordinated to other nodes having other types of specialization.

A technique aimed at generalizing the characteristics of specialization has been developed. Cities are grouped by levels and types of specialization, creating a multidimensional functional classification.

The method used to determine the profile of cities

Readers whose interest centres predominantly on the substance of Soviet problems, rather than on econometric method, can jump over this section and the next. It will be possible to follow the reasoning without turning to explanations of method.

The initial data used for processing form a matrix of 221 × 8. Each element in the matrix is represented by a positive integer a_{ij}, indicating the number of people who in the ith city make up the work force of the jth economic sector ($i = 1, 2, \ldots, 221$); $j = 1, 2, \ldots, 8$). The first operations involving a_{ij} are designed to ensure comparability among cities, and then among sectors.

First, the impact of city size on specialization must be eliminated. This elimination is accomplished by dividing each a_{ij} by the sum of the corresponding row. We thus get $x_{ij} = a_{ij}/a_i$. The significance of each sector is thus expressed in fractions. It might equally well be expressed in the form of a percentage of total employment in the given city. The overall magnitude of employment in the city can thus be ignored.

Second, the eight employment sectors must be given equal weight in the way they differentiate the cities. This balancing can be achieved by eliminating the impact of differences in the magnitude of employment, which are particularly noticeable, for example when the size of the labour force in manufacturing and government administration are compared. Because of the different nature of work in these two sectors, one (manufacturing) generally accounts for tens of per cent of overall city employment while the other (administration) accounts for fractions of one per cent. The power of resolution of our typology must be the same for both the tens of percentage in the first case and the percentage fractions in the second case. A separate step is therefore needed to equalize the attention given to each employment sector: the columns X_{ij} must be statistically standardized.

A yardstick is needed to measure each employment sector with regard to variation observed in employment sectors across all cities. Mathematical statistics suggests the most suitable measure to be applied as such a yardstick: the standard deviation for the columns X_{ij}. The standardization procedure appears as follows, in BASIC notation:

$$X(ij) = (x(ij)-AVERAGE(j))/SQR(VARIANCE(j)). \quad (2.1)$$

The next step is to sort the economic sectors according to three levels of significance: satisfaction of local needs, supply within a regional radius, and long-distance hauls outside the region. The boundaries between these significance levels were established on frequency graphs constructed for each sector of employment (see Figure 2.1).

The frequency graphs plot class intervals for percentage of employment in a sector (x-axis) with the number of cities (y-axis) that occur in the class intervals. It is remarkable that all eight employment sectors show similar positions of breakpoint on the curve.

First, the curves change direction around the mean value of employment percentages projected on the abscissa. Second, there are sharp breaks in the slope near the points at plus or minus half of the standard deviation from the mean for all projected values on the x-axis.

The coincidence of the breakpoint in the frequency graphs (Figure 2.1) is not accidental. It originates in the geometry of the network of settlement. Percentages displayed by the graphs, certainly, have economic and technological manifestations, but at the same time they depend on externalities of location. The size and density of the nearest neighbour places is a factor of location, with similar impact on all activities in a city. It is a benefit for every activity if the spacing of other cities permits easy co-operation and access to the combined labour market. But if there is too much crowding (as in the Donbass, for example), then difficulties with water supply and environmental pollution cut down growth in all activities. This peak in the development process accounts for similarities in the frequency graphs of all the activities. The similarity allows us to adopt identical rules for determining the mean-

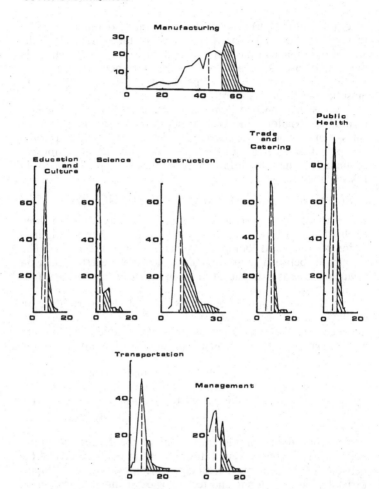

Figure 2.1 Frequency distributions of the percentage of employment in each of eight economic sectors for 221 cities of the USSR (population 100,000 or more in 1970). Percentages of employment are shown on X axis; number of cities on Y axis. Cities with long-haul links are shaded.

ing of different segments of curves displayed as in Figure 2.1.

As we see it, the breakpoint of the graphs have the following interpretations:

1. If the employment share of a particular sector falls below the mean, it can be said to meet mainly local needs.
2. If it falls within the interval between the mean and plus or minus half of the standard deviation, the sector meets regional needs.
3. If the employment share is higher then half of a standard deviation above the mean, the sector is dependent upon long-distance links.

Obviously, it would be wrong to claim that these statements are exact descriptions of real-life relations in any city, taken individually. Because they describe a statistical trend, applicable, as a rule, only to groups of cities, conclusions based on them must also be addressed to aggregations of urban places rather than to individual centres.

Hereafter the three levels of significance of sectoral employment are interpreted as local, regional, and long-haul (or distant) transportation flows for groups of cities under consideration. We shall designate a numerical code to levels of specialization (with accompanying transporation flows): 3 – local; 2 – regional; 1 – long-haul. The variables x_{ij} can thus be replaced by the numerical code.

Information about all the sectors represented in each city can now be written as an eight-digit code. For this purpose each activity must get a fixed place within an objectively designated code. The first digit of the code represents the level of specialization achieved by a particular city's manufacturing sector; the second represents the level of its construction sector, and so on, as follows:

1 – Manufacturing
2 – Construction
3 – Railways and other transportation
4 – Trade
5 – Health
6 – Education and culture
7 – R&D ('science')
8 – State economic management and administration

For example, the numerical code (13333333) represents a city where manufacturing contributes long-distance freight for railways and all other activities are for local needs only.

The point of the code is to distinguish city activities in terms of their ability to generate probable links with a radius which crosses boundaries of Soviet economic regions. Very long hauls in Soviet conditions take the form of long-distance flows on rail mainlines.

Railway flows are designated by the symbols '1', '2' or '3' in the third place in our eight-digit codes of cities. Freight may go in both directions: first, as supply; and second, as products of activities. A partial exception is the construction industry. Many of its creations are obviously immobile: the structures of huge hydroelectric projects, industrial plants, etc. In this case the numerical code represents the radius over which building materials and equipment are supplied to construction sites. This exception must be borne in mind in any interpretation of the findings. Nor is the entire construction industry an exception. Factories producing construction materials and building modules often serve their zones of influence by shipping finished or near finished products on platforms of freight trains.

There are other ways of interpreting employment in the construction sector. For example, the making of building materials may be partly attributed to another sector (manufacturing); there are wide differences in the character of products in terms of transport haul: cement is a building material which circulates in inter-regional trade, while bricks and building blocks are strictly for local use. Furthermore, labour input in relation to tonnes of supply is likely to vary widely depending on whether a particular project is in making buildings or tuning the equipment in them. Despite all these qualifications, we thought it useful to treat the construction industry as a separate sector of employment. In employment percentages it usually ranks second, after manufacturing.

The method of determining types of cities

Our task now is to highlight similarities among 221 numerical codes, and later to sort them into groups. Algorithms of numerical taxonomy show us how to approach the job.

The groups are, in essence, fuzzy sets of the eight-digit codes. The codes may claim acceptance into the same set if all their eight elements are identical, and also in cases of specifically permitted disagreements for one or several elements. It is easy to program a computer for steps comparing the code: upon subtracting one code from another one adds together, disregarding the sign, all elements in places 1 to 8 of the resulting 'dissimilarity vector'.

It is a matter of judgement how fuzzy to make meaningful groups of cities. Initially we set the following limit of fuzziness: if the difference is equal to or greater than 2, the compared codes are not placed in the same group.

Each has been compared with all the others to find all possible pairs that could be formed within the overall set of 221 codes. One cycle of comparisons means calculating the differences between codes in $221!/(2!219!) = 24{,}310$ combinations. The comparison cycles follow several times, one after another. They bring city groupings, first, with identical codes, then with differences in the level of specialization within one activity, within two activities, and so on.

The results lead to the construction of a tree diagram of similarity. Each stage in the coalescing of branches means that cities shown in the form of branches are considered similar.

The notion of similarity is interpreted increasingly liberally from stage to stage. At first only cities with absolutely identical eight-digit codes are combined. Then we add those displaying a slightly different level of specialization in one sector, then in two sectors and so on.

The clustering algorithm in use makes it possible to identify, first, the groups that are known as single-linkage phenons in numerical taxonomy. Then groups that have the characteristics of completely linked phenons are identified (Baily, 1970; James, 1985). The difference is that, in the first case, cities making up the core of the group are joined by those in which the level of specialization deviates by one level in either direction from the level observed in the core. The group thus accepts some cities that among themselves are more different than permitted in the paired comparisons performed with the core code. The second case allows deviations from the core code in one direction only, ensuring strict similarity within the group of cities. One cannot start directly with the task of identifying the linked phenons, because many trials are required to find core codes right in the middle of the range permitted for groups.

The criterion of similarity among cities is the taxonomic distance computed in respect to elements of the eight-digit codes:

$$D = \sum_{i=1}^{n=8} |K_i - k_i| \qquad (2.2)$$

where D refers to taxonomic distance, n to the digit position within the eight-digit code, K_i, k_i are the corresponding elements from codes of the compared cities, with $|K_i - k_i|$ always being less than 2. The code K_i characterizes a city that makes up the core of the group. The grouping stages lead to a tree diagram, branching at D

= 0, 1, 2, 3, 4. Further than $D = 4$ we do not move because it would yield groups that are too small.

To create a grouping that does not require calculation of D, another approach is also used. The use of D assumes identical significance in the level of specialization for each of the eight sectors of employment. But another point of view also demands attention, namely one that sees a significant difference between secondary and tertiary activities. The material production (or secondary) activities project to the first four digits in the code, and the tertiary ones to the last four digits.

The grouping of cities according to this alternative approach results in the graph shown in Figure 2.2. Each type of city is plotted in the *x, y* co-ordinate space, in which

$$x_j = \sum_{i=5}^{n=8} k_{ij} - \sum_{i=1}^{n=4} k_{ij}; \qquad y_j = \sum_{i=1}^{n=8} k_{ij} \qquad (2.3)$$

In this case $k_{ij} = 1$ is a symbol for *i*th city if its *j*th sector is large enough to require long-haul links.

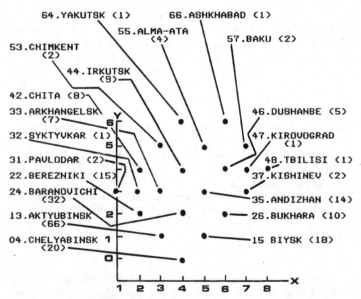

Figure 2.2 Soviet cities distributed according to the dominance of material production (X axis) and dependence on the long-haul links (Y axis).

70

The dominance of either the secondary or tertiary sector (on the level which calls for long-distance links) is plotted on the y-axis. The y-axis thus distinguishes cities in terms of the number of their links that extend beyond the given region. In the (x, y) co-ordinate space, the entire sample of 221 cities falls into eleven groups, which could be viewed as points or cells separated by taxonomic distances.

For a more generalized image of Soviet cities we find that eleven groups reduce to five or even three groups. In the last two cases, the grouping would be designated in respect to either the x-scale or the y-scale, but not both.

Uncovered types of cities

The set of levels of specialization, expressed for each city by the eight-digit code, makes it possible to define groups of cities of particular types. The most distinctive group has a core made up of cities with the code (13333333). This absolutely identical profile, indicating prominence of manufacturing with low content of all other activities, belong to twenty-one cities. Typical representatives of this category of cities would be Tula and Ivanovo.

Another clear-cut type of city is characterized by prominence of two different activities, with substantially equal levels of significance, as illustrated by Riga and Vladimir. A distinctive group is formed by cities specializing in science and management, with the core made up of Moscow and Kiev.

By using the (x, y) co-ordinate space displayed in Figure 2.2, it is possible to define the typology of cities in terms of the number of prominent activities (each demanding long-distance links) and also in terms of balance between material production and tertiary or information-related activities. There are groups that could be accommodated in the cells of a 7×8 matrix, in which the cells corresponded to integers on the x and y axes. A total of twenty-one cells in the matrix were filled.

As a first approximation, the 221 cities were broken down into twenty-one groups. Such a number of types for cities is clearly too large, because some of the groups consist of only one or two cities. But the listing of twenty-one groups is useful in pointing out extremes in the characteristics of a number of cities. For this purpose consult Appendix 2.1 where the distribution of the 221 cities among all twenty-one groups of urban specialization is shown.

The distance between groups can be judged from the scales on the x and y axes, in Figure 2.2. Each group is designated by two

71

integers, which can be expressed as positive numbers, i.e., in the form of indices for the cells in the 7×8 matrix. The first digit in the index gives the number of activities in the city that are endowed with long-distance links – a maximum of 6 and a minimum of 0. The second digit indicates the preponderance of either secondary or tertiary activities, with the extremes represented by 1 and 8. Equilibrium between the material production activity and the tertiary activity positions cities in the cell at $x = 4$.

When twenty-one groups are distinguished, a special place is occupied by those that fall in the uppermost cells of the matrix (Figure 2.2). These cities are unique. The lack of similarity is shown by the cell indices, as follows:

64 – Yakutsk
66 – Ashkhabad
47 – Kirovograd
48 – Tbilisi
32 – Syktyvkar

Four other groups consist of two cities each:

53 – Chimkent, Gur'ev
57 – Baku, Tashkent
31 – Pavlodar, Tyumen
37 – Kishinev, Ordzhonikidze

The number of cities in the remaining twelve groups increases from the upper to the lower row cells in the matrix. There is also an element of growth in the columns from left to right. The group indices are as follows: 55, 42, 44, 46, 33, 35, 22, 26, 15, 04, 13, 24. There is a tendency for contrasts between groups to decline as the groups become larger.

Cities that form part of a larger group tend to be located in the interior of the USSR and within the main belt of settlements. It would seem that the dense clustering of cities fosters development of similarity traits. The opposite impact seems to be produced in a more wide-meshed network of cities as well as in peripheral city locations. These two situations usually coincide. They foster an increase in the number of city activities with long-distance links.

By combining the extreme columns and the three uppermost rows in the 7×8 matrix (Figure 2.2), one can eliminate the groups with small numbers of cities. As a result, the 221 cities fall into eleven groups. Such aggregated groups retain their objective basis. The similarity among cities within each of the eleven groups corresponds to their co-ordinates in the x, y space.

The final results of our urban typology are shown in Table 2.2.

One may compare the number of cities that fall into each of the eleven groups in terms of their *x, y* co-ordinates. Table 2.2 is constructed to provide a two-fold classification. A more generalized classification can be achieved by adding up all the cities by either rows or columns.

By adding up the columns, we find that the 221 cities of the USSR form three networks. In 101 cities, secondary economic specializations predominate, and they form the first and densest network. Another network is made up of sixty-two cities in which secondary and tertiary activities are balanced.

The least dense network, made up of fifty-eight cities, displays predominantly tertiary specializations. This group includes most of the republic capitals, including Moscow. The third network is distinguished by a far more dispersed distribution than is observed in the whole set of 221 cities.

Another simplification in typology is achieved by adding up the rows in Table 2.2. This process yields four networks. The densest of these is formed by eighty-four cities in which only one activity is endowed with long-distance links. The next group includes the sixty cities in the top row of Table 2.2 where at least three activities have long-distance links. The third group, nearly as large as the second, with fifty-seven cities, includes those that have two highly pronounced specializations. And the least dense of the groups includes twenty cities that are internally balanced and have no activities that make themselves felt outside the region or the local zone of influence.

Table 2.2 Functional typology of Soviet cities

Number of activities with long-distance hauls (y-axis)	Relationship between secondary and tertiary activities (x-axis)					
	Material production		Balance or near it	Tertiary activities		All cases
	Pronounced	Moderate	cases	Moderate	Pronounced	
3	11	9	10	18	12	60
2	15	–	32	–	10	57
1	–	66	–	18	–	84
0	–	–	20	–	–	20
Total	26	75	62	36	22	221
Position on x-axis (Figure 2.2)	−3, −2	−1	0	+1	+2, +3	

Do the cities of the USSR display a relationship between the two dimensions that serve as the basis for classification in Table 2.2? In other words, does the predominance of tertiary activities grow in a city as the number of activities with long-distances links rises?

To answer that question, one should test the statistical hypothesis that the frequencies shown in the eleven cells of Table 2.2 were similar to those that would have occurred solely on the basis of the sums of its rows and columns. When frequencies are equal to the product of a row sum ($m_{i.}$) and a column sum ($m_{.j}$) divided by the total sum of frequencies ($m_{..}$),

$$m_{ij} = (m_{i.} * m_{.j})/m_{..} \qquad (2.4)$$

then it shows the independence of the two bases in a two-way classification table. If the number of cities in the eleven groups of Table 2.2 is designated as the matrix M_1, then the computations on the basis of formula (2.4) would lead to matrix M_2. The null hypothesis used for testing takes the form H_0: $M_1 = M_2$.

Such a test is known in statistical analysis as an inference concerning proportions in contingency tables (Kilpatrick, 1987: 21–15). Because of mandatory empty cells in Table 2.1, there is a low number of degrees of freedom (DF = 3) for computed elements of M_2. A comparison between M_1 and M_2 shows that the null hypothesis must be rejected. The value of the chi-square criterion calculated for M_1 and M_2 is almost twice as large as the tabulated value for a 1 per cent level of significance ($\chi^2 = 25.45$; $\chi_{tab}^2 = 13.277$).

Thus, a growth in the number of activities with long-distance links, resulting from a national division of labour, tends to be associated in Soviet cities with an increasing presence of tertiary activities. Of course, such a presence does not necessarily mean that tertiary activities are the leading sector of employment; we only look here at the relative significance of activities and their ability to generate longer links. Furthermore our conclusion is based on the structure of employment as of the single year 1970. If we assume that the evolution of events through space is similar to their evolution through time, then the conclusion can also be used for predictive purposes, namely that tertiary activities are the ones most likely to generate longer links in a city after the number of activities with a nationwide specialization reaches two or more. This finding, in turn, suggests that new specializations are most likely to be tertiary.

By statistical testing of hypotheses, one is also able to obtain a clear-cut answer to the question whether there was, in 1970, a difference between older populous cities and those that grew to substantial size only during Soviet five-year plans. The distribution of the 221 cities among eleven groups shown in Table 2.2 was compared with the distribution of the sixty-six cities that had a population of at least 70,000 in the 1926 census. In this case the $H_0 : M_1 = M_2$ was rejected.

The basis for rejecting the hypothesis was the chi-square value $\chi^2 = 21.64$, which exceeded the tabulated value of $\chi_{2tab} = 13.28$. With a low degree of probability – one chance out of a hundred is clearly excessive – we thus find no similarity between old and new cities. In the structure of specialization and the way the magnitudes *x, y* are expressed, the entire set of 221 cities looks quite different from the subset of sixty-six cities.

Comparing the old and new cities in terms of their allocation to the eleven groups, one may discover the following:

1. Among the sixty-six cities that were populous in 1926, there was clearly a sharp predominance of situations with a pronounced bias towards tertiary specializations ($x > 2$) or with a less pronounced dominance of such activities ($x = 1$); there was also a high frequency of situations in which long-distance links were altogether absent ($y = 0$). The latter characteristic would normally be associated with cities situated within the main belt of settlement and near other major centres.
2. New cities exhibit a higher frequency of bias towards economic activities ($x < -2$) combined with a higher frequency of long-distance links. This category includes manufacturing centres that have arisen in backward peripheral locations.

A complete list of all studied cities according to their resulting types appears in Appendix 2.1.

Networks of specialized cities

Location patterns of Soviet urban centres which have activities related to long-distance transportation are displayed in Figures 2.3 and 2.4. The networks in each case reflect a peculiar geometry inherent in the entire urban system of the USSR. Cities do not fill the territorial space of the Soviet Union in any kind of uniform pattern. They tend to concentrate within the main belt of settlement, forming a wedge that has its broad base in the west (between Odessa and Leningrad) and comes to a point near Novosibirsk, along the Trans-Siberian Railway. A few clusters of cities occur

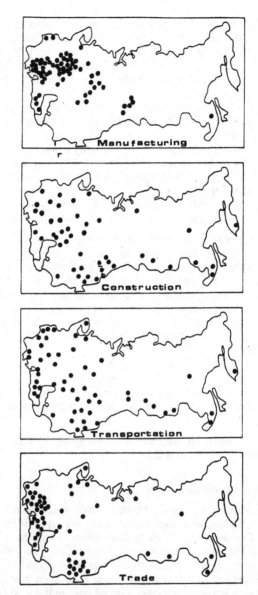

Figure 2.3 Networks of cities with long-haul links: (a) manufacturing; (b) construction; (c) transportation; (d) trade.

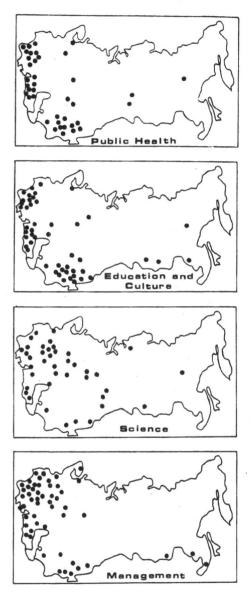

Figure 2.4 Networks of cities with long-haul links: (e) public health; (f) education and culture; (g) science; (h) management and administration.

beyond the wedge, in the valleys of Transcaucasia, in the oases of Central Asia, in the Kuznetsk Basin, etc. The overall geometry of the network of cities is not always duplicated in the eight networks for which characteristics are shown in Table 2.2. The distinctive nature of each of the eight activities leaves an imprint on the corresponding network of cities.

One way of contrasting the eight long-distance networks might be to compare the frequency graphs shown in Figure 2.1. Three categories of curves can be distinguished. A unimodal frequency distribution is found for five activities: construction, transport, trade, health care and education. A bimodal distribution is reflected in science and management, and a multimodal distribution applies to manufacturing. The presence of several maxima in the frequency graph suggests a pronounced heterogeneity of the objects.

In manufacturing, there is in fact plenty of evidence of internal distinctions: location criteria vary for each branch of industry. For the seven other activities, differences in location requirements are evident from differences in the slope of the curves. The steeper the slope, the greater are the differences in significance of the particular activity in various cities, and the more sensitive is the particular activity to the resource situation or to proximity of other economic centres. Centres at similar hierachical levels fall into segments of the curve between two breakpoints. Even though the two basic breakpoints on the curves generally coincide, there are still quite a few distinctive features in each activity.

The peculiarities evident in the frequency graphs of Figure 2.1 are associated mainly with the geometry of the city networks and with hierarchical features expressed in the nearest-neighbour location of cities. Evidence of the command-type economy of the Soviet Union does not appear on the frequency graphs. The curves in Figure 2.1 turned out to be quite similar to curves constructed for cities of the United States (Nelson, 1955). The fundamental differences between the USSR and the USA in their social systems and methods of economic management would rule out any similarity in the field of economic phenomena. But when it comes to the geometry of city networks, the two countries have a great deal in common: the size of their territory, the sequence of natural zones, etc.

With a view to identifying the varying geometries of networks made up by cities with long-distance links, we compiled the eight maps shown in Figure 2.3 and 2.4. One major geographical fact becomes evident. Except for the centres of manufacturing and science, all the networks appear to be made up of cities that are well outside the main belt of settlement.

The purely geometrical properties of two-dimensional space clearly affect the character of networks made up of cities with a highly pronounced specialization. Other things being equal, any peripheral point would obviously have a longer radius of links with the national territory than with any given point in the interior. Therefore the mere presence of a particular specialization in a peripheral city would make it highly unlikely for another city of similar specialization to be situated nearby.

What we are witnessing is a deepening of specialization, rather than a levelling such as might occur if incremental growth of a particular activity were to be spread among several cities close to one another. In general, therefore, there is a high probability for greater specialization to occur on the periphery.

Population size is another factor influencing whether a particular long-distance specialization will arise in a city. The mean 1970 population of cities in the eight networks ranged from 799,000 in the science network to 188,000 in the construction network, or a ratio of 4:1. This finding supports the proposition that science activities gravitate to more mature urban centres while the building industry is focused in places where urbanization is still in its early stages.

A larger population size would also be expected for centres prominent in State economic management and in administrative functions. Such cities include the capitals of major civil divisions or of union republics. The mean population of such administrative centres in 1970 was 419,000. The Soviet Union, like Imperial Russia before it, gladly displays symbols of State power in the most populated urban centres, both at national and regional levels. In this regard it is in sharp contrast with practices of the United States, where economic and political capitals do not coincide on the national level and rarely do on the regional level.

For manufacturing, trade, transport, health care, education and culture, the cities with long-distance links display similar mean populations, ranging from 227,000 to 288,000 in 1970.

The highest level of specialization does not necessarily occur in the largest cities. Only science and, to a much lesser degree, management encompass several million-population cities among those where these activities are marked by long-distance links. Most of the cities with such a high level of specialization have 100,000 to 500,000 residents.

How legacy of industrialization determines growth of cities

Features of Soviet cities captured by this typology originate from five-year plans of industrialization. Table 2.2 displays it very clearly. It is interesting to test the extent to which positioning of cities in the cells of Table 2.2 (i.e. legacy of industrialization) predetermines their growth rate up to the present time. This test is equivalent to checking how strongly present Soviet cities are chained to the dated legacy.

Appendix 2.1 permits us to determine median percentages of growth for each of eleven types of Soviet cities. The percentages are shown in Table 2.3.

Table 2.3 Median growth for uncovered types of Soviet cities: 1970–86 (1970 = 100 per cent)

y-dimension of typology	x-dimension of typology				
	0	*1*	*2*	*3*	*4*
3	63.6	48.0	47.0	49.0	50.0
2	21.0	—	33.6	—	43.2
1	—	25.7	—	42.5	—
0	—	—	39.3	—	—

For each cell of Table 2.3 there is a city which in 1970-86 had a median growth among all other similar centres. The similarity is equivalent to attributing identical values from x- and y-axes to the cities. Remember that the axes have marks which tell what the cities are in balance between material production and tertiary activities (x-dimension), and in the number of sectors that demand for long-distance hauls.

Cell (2,0) of Table 2.3 is exemplified by Ufa. This city had 39.3 per cent growth in 1970-86. This is a middle position in a list of twenty cities similar to Ufa. In the same way all other cells of Table 2.3 have specific examples:

    (1,1) – Rostov/Don from a group of 66 similar cities
    (3,1) – Ryasan'   from a group of 18 similar cities
    (0,2) – Novosibirsk from a group of 15 similar cities
    (2,2) – Tambov   from a group of 32 similar cities
    (4,2) – Frunze   from a group of 10 similar cities
    (0,3) – Tyumen'   from a group of 11 similar cities
    (1,3) – Bel'tsy   from a group of 9 similar cities
    (2,3) – Uralsk   from a group of 10 similar cities
    (3,3) – Alma-Ata   from a group of 18 similar cities
    (4,3) – Tashkent   from a group of 12 similar cities

Plate 1. Near the centre of Moscow: a mix of industries and apartment houses.

Plate 2. Typical houses for any Soviet city: the style of Khrushchev's decade, the 1950s.

Plate 3. A modern hotel (Dmitrov Street, Moscow): for the privileged party bureaucrats only.

Plate 4a. Inhabitants of the Soviet cities.

Plate 4b.   Inhabitants of the Soviet cities.

Plate 4c.   Inhabitants of the Soviet cities.

Plate 5. A new generation of urban dwellers guided by their teacher.

Plate 6. One of the Soviet gate-keepers: at the entrance to the French Embassy in Moscow.

Plate 7. The new generation: buildings and people.

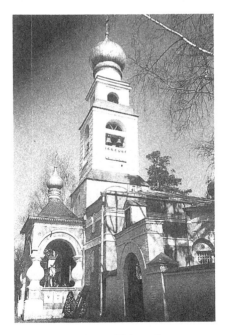

Plate 8. Relics of Old Russia.

Plate 9. A typical decoration for a Soviet city: Lenin's monument.

Plate 10a. Public transportation for urban residents.

Plate 10b. Public transportation for urban residents.

Plate 11. Railways to provide access to suburbs.

Plate 12. A Moscow suburb: neglected and famous (Pasternak's house).

Plate 13. A more common suburban residence in Central Russia.

The sum for all cases is 221: all cities in the studied sample.

Because the content of typology is in (x, y) values it is quite logical to check how successfully such values predict the actual growth rates displayed in Table 2.3. We apply regression analysis for that.

From examining the entries in Table 2.3 it is clear that the growth rate increases start from its south-western corner. The gradient of increase goes generally towards the north-eastern corner, and it is a non-linear, because the steepest increase occurs along the left margin of Table 2.3 (up to the rates of growth in Tyumen'). With entries for just eleven types of cities, one must operate with the least possible number of independent variables in order to have distinct results from regression analysis. Many curving surfaces are excluded in fitting a regression formula to data points of Table 2.3.

A reasonable approach is, naturally, to merge (x) and (y) for calculating distance in the taxonomic space from a pole in it, which is $(0,0)$ of Table 2.2. In BASIC notation such distance is:

$$D = SQR (x^2 + y^2). \tag{2.5}$$

If rates of urban growth in Table 2.3 are designated as $ROUG(x, y)$ then one of our successful tests with a non-linear formula may be written as:

$$ROUG(x, y) = 21.236\ D^{0.588}$$

This power function explains 50 per cent of all variances in actually observed $ROUG(x, y)$. A much better prediction of urban growth is, however, possible.

Let us rearrange all $ROUG(x, y)$ from Table 2.3 in decreasing order of numerical values:

$ROUG(x, y)$	D	
63.6	2.83	
50.0	7.21	*
49.0	5.66	*
48.0	2.83	*
47.0	4.24	*
43.2	3.16	*
42.5	2.00	*
39.3	1.41	*
33.6	1.44	*
25.7	0.0	*

The parallel column of D is computed starting from the pole

located at (1,1) cell of Table 2.2. It is a better positioning than the initial test with its (0,0) location of the pole: one must keep in mind that we have freedom to do it because the zero level in the scales of taxonomy is a matter of convention. Next, if we observe the ranked values of ROUG(x, y) paralleled by D, it is clear that all departures from a linear function are very local: they are just at the top of the series. This is a case of spurious behaviour at the extremes of the studied conditions.

If there is spurious behaviour at the borders of the range for observed growth rates, then the consensus in statistical analysis is that such values should be excluded. In all other values in the starred rows above, there is very little non-linearity, and we get remarkable success with the regression formula:

$$ROUG(x, y) = 32.754 + 2.99 \, D \qquad (2.6)$$

$$(R^{\hat{}} 2 = 0.712; \; F = 17.35)$$

Sometimes in the hard sciences the quality of the formulae like (2.6) is assessed by consulting the tables of F at its 1 per cent point distribution. It signals adequacy of a regression, roughly speaking, with 99 per cent certainty. In our case such a rigorous test gives 16.26 as tabulated F, and we are above it. In the social sciences, where definitions are often fuzzy, it is common to consider more relaxed standards for F: usually from 5 per cent distribution of F. This would mean F(tab) = 6.61, and we are 2.6 times above it. That is our safety margin over 95 per cent certainty in claiming that a regression response is distinct and clear.

All this proves the following: Soviet urban growth between 1970 and 1986 has been conditioned by the legacy of industrialization. There is a dominance of the legacy, because the coefficient of determination ($R^{\hat{}} 2 = 0.712$) is on the level, which permits one to say: more than 70 per cent of the variance of urban growth is a response to the legacy of industrialization.

Really, Soviet cities are chained to that legacy. Any plans to move them into the post-industrial phase are likely to be terribly difficult to implement.

The impact of industrial cities on railway flows

In the most obvious form, the phenomenon of specialization is represented by those activities that are designated by the symbol '1' in the eight-digit code. Such activities require long-distance links, because they serve a large number of places beyond the boundaries of the particular region. Considering the size of the

Soviet territory, it is unlikely that trucks can compete with trains in carrying freight over long distances.

To illustrate, here are some of the more important train distances and times (in the ordering suggested by the gravitation model of urban interaction):

Moscow to Leningrad (650 km)   6–8 hours
Moscow to Kiev (872 km)   12–13 hours
Moscow to Riga (922 km)   14–15 hours
Moscow to Sverdlovsk
(1,818 km)   27 hours
Moscow to Baku (2,475 km)   43 hours
Moscow to Novosibirsk
(3,343 km)   48 hours
Moscow to Irkutsk (5,191 km)   81 hours
Moscow to Vladivostok
(9,297 km)   151 hours

In our sample of 221 cities there are 432 cases of intensive specialization, and they all contribute to the present overloads of Soviet railways. Most of the enterprises existing in 1970 have not ceased to operate; on the contrary, it is more likely that in recent years they have increased in size. If new local sub-contractors have replaced some distant supply, it is likely that the sub-contractors, in their turn, demand or generate distant flows, at least in a volume comparable to that of 1970.

Manufacturing dominates in terms of the number of cities generating long-distance links. It is followed, with almost as many cities, by transport, trade and management (57 and 59). Thinner

Table 2.4 Number of cities forming networks with long-distance freight and their population of 1970 and 1986 (numbers in brackets are for 1970)

Activities which demand long-distance freight hauls	Threshold of acceptance, employment share (%)	Cities forming the network		
		Number of all cities	Population over 1 million	size 500,000 to 1 million
Manufacturing	(50.3)	(71)	3 (3)	16 (5)
Construction	(3.1)	(48)	0 (0)	1 (1)
Transportation	(9.5)	(59)	1 (1)	4 (3)
Trade	(7.6)	(59)	1 (1)	2 (2)

Note: The following specialized activities have negligible participation in freight flows (activities are shown with corresponding 1970 values for columns A, B, C, D: Health (6.0, 46, 2, 3); Education and culture (7.7, 52, 2, 5); R & D (4.5, 40, 9, 8); Management and administration (6.5, 57, 4, 3).

Soviet Urbanization

Table 2.5 Legacy of urbanization in the estimated long-distance freight for Soviet railways (1.1.1986) Samples by Economic Regions

Regions	Centre contributing to freight (number of activities, if > 1)	Sums of W1*W2
	Western fringe of the Soviet Union	
Baltic	Tallin (2), Kaliningrad (2) Klaipeda (2), Daugavpils (2)	2,242
Belorussia	Mogilev (2), Grodno (2), Brest (3), Baranovichi (2) Orsha (2)	2,448
South-west	Vinnitsa (2), Cherkassy Khmelnitsk (3), Belaya Tserkov' (2)	2,073
Donets-Dniepr	Donetsk (2), Zaporozhye, Zhdanov, Stakhanov (2) Kharkov (2), Poltava, Voroshilovgrad, Gorlovka, Makeevka, Dneprodzerzhinsk, Kremenchug, Kramatorsk, Melitopol', Nikopol', Slavyansk (2), Berdyansk (2), Kommunarsk (2), Lisichansk, Konstantinovka, Kr. Luch (2)	10,973
South	Kherson, Kerch, Simpheropol (4)	1,854
Moldavia	Bel'tsy (3)	453
	Central core of the Soviet Union	
Central	Yaroslavl, Tula, Ivanovo, Bryansk, Lipetsk, Kostroma (2), Andropov, Podolsk (2), Lubertsy (2), Kolomna (2), Mytyschi (2), Kovrov, Elektrostal', Serpukhov, Orekhovo-Zuevo, Noginsk	5,394
C. Chernozem	Belgorod, Tambov (2) Yelets (2)	1,132
North-west (as before 1983)	Arkhangelsk (2), Murmansk (4), Cherepovets, Vologda (4), Petrozavodsk (4), Novgorod, Severodvinsk (3), Syktyvkar (3)	6,957
Volga-Vyatka	Gorky (2), Dserzhinsk (2), Cheboksary, Yoshkar-Ola (2), Kirov, Saransk	4,976
Povolz'ye	Kuybyshev, Kazan (2), Saratov (2), Togliatti, Ulyanovsk, Penza (2), Volzhsky (2), Balakovo, Syzran, Novokuybyshev	8,403
N. Caucasus	Rostov, Krasnodar (2), Sochi (4), Taganrog (1), Shakhty, Novocherkask, (2), Novorossiysk (2), Armavir, Novoshakhtinsk	4,992
	Southern fringe	
Transcaucasia	Sumgait (2), Sukhumi (4), Leninakan (2), Kirovokan	1,587
Central Asia	Chirchik (2), Kokand (4), Osh (4)	1,804

84

Table 2.5 Continued

Kazakhstan	Aktuybinsk, Pavlodar (3), Karaganda (2), Temirtau, Chimkent (5), Dzhambul (4), Semipalatinsk, Gur'yev (5), Ust-Kamen. (2), Uralsk (5), Kustanay (4), Petropavlovsk. Tselinograd (4)	10,442
	East of the Soviet Union	
Urals	Berezniki (2), Kamensk-Ur., Kopeysk, Magnitogorsk, Miass, Nizh.Tagil, Orsk (2), Orenburg, Pervouralsk, Serov (2), Slatoust, Sterlitamak, Ustinov (Izhevsk)	4,350
W. Siberia	Anzhero-Sudzhensk, Barnaul, Belovo, Kiselevsk (2), Leninsk-Kuznetskiy, Novosibirsk (2), Prokopyevsk (2), Rubtsovsk, Tyumen' (3)	3,552
E. Siberia	Angarsk, Bratsk (2) Chita (4), Irkutsk (4), Norilsk (3), Ulan-Ude (2)	5,748
Far East	Khabarovsk (4), Komsomolsk/Am., Nahkodka (2), Petropavlovsk/Kam. (4), Vladivostok (2), Ussuriysk (3), Yuzhno-Sakhalinsk (4), Yakutsk (6)	
	Total	86,762

long-distance networks are found in cities with a high level of construction, health care and scientific research activities. In general, network density tends to be higher for long-established activities.

Soviet statistical yearbooks ('Narkhoz') permit one to know the total count of rail freight: 2,896 million tonnes in 1970 and 4,078 million tonnes in 1986. But they are silent about the regional distribution of the freight, which of course would suggest the location of bottlenecks on the Soviet rail system. A certain amount of inferential reasoning, however, permits one to learn what, in the main, shapes Soviet rail freight, which is rooted in the legacy of industrialization. The 1970 picture, shown in maps of specialized cities (Figures 2.3 and 2.4) can be updated with information from Table 2.4.

We may assume that each of the Soviet economic regions has a share in long-distance freight. That is likely to be proportional to the number of specialized functions of material production in its cities. This gives the first weight coefficient (W1) to improve the initial assumption.

Next, the population size of a city may serve as another weight coefficient (W2) because it indicates the scale of all operations in an urban centre. By multiplying W1 and W2 one may arrive at a better estimate of distribution of the freight among the regions. It is reasonable to apply W1 and W2 in calculations to all cities of the studied sample that emphasize material production, as it has been uncovered by the typology. Meaningful proportions may transpire, of course, only for groups of cities, and for this reason we aggregate sums of (W1*W2) for each of the economic regions of the USSR in Table 2.5. These sums illustrate the burdens created in each economic region by activities that are outdated.

The estimated sums of W1*W2 in Table 2.4 may be calibrated by assuming that TOTSUM (W1*W2) = 86,762 is directly proportional to the officially reported total rail freight in the Soviet Union: 3,951.2 million tonnes in 1985. We do not know the coefficient of proportionality (k), but as a rough guess it may be set to unity. We do assume that: (a) Soviet railroads serve only long-distance hauls; and (b) our samples in Table 2.5 adequately cover all such hauls.

The equation

$$86,762 * k = 3,951.2 \text{ million tonnes} \tag{2.7}$$

permits us to determine $k = 45,540.7$. This is the number of tonnes which may be attributed to each unit displayed in the right-hand column of Table 2.5. These estimates are all very approximate, of course. Our calculations may be good, perhaps, only as an attempt to determine the ordering of Soviet economic regions in the burden imposed on them from the dated legacy of industrialization.

With all our reservations, it remains possible to think that the following ordering of regions is correct:

1. Donets-Dniepr
2. Kazakhstan   (2,080.5)
3. Povolzhye
4. Far East
5. North-West
6. Eastern Siberia
7. Central Region
8. Northern Caucasus
9. Volga-Viaytka
10. Urals
11. Western Siberia
12. Belorussia   (786.9)
13. Baltic Region   (760.7)
14. South-West

15. South
16. Central Asia   (1,932.4)
17. Transcaucasia   (1,142.7)
18. Central Chernozem
19. Moldavia   (311.5)

To validate the suggested ordering there are several entries which are indicated in parentheses above. These figures were taken from the Soviet Statistical Yearbook of 1986; they represent million tonnes of freight carried in 1985 by trucks. Because officially reported figures are aggregated for all the Soviet Union's republics, we are left with rather partial possibilities for comparisons.

There are only two regions that do not fit exactly in the suggested ordering: Central Asia and Transcaucasia have bigger figures of the freight carried by trucks, than our estimates for the railways may suggest. But both of those regions also have reasons for carrying more freight by trucks in relation to rail freight. Central Asia and Transcaucasia are regions with gigantic cotton plantations, and the plantations rely mostly on truck deliveries. It happens because the rail network is inadequate for serving the plantations. Mountain terrain limits the development of rail networks in both regions. However, Transcaucasia and Central Asia have the same ordering in our estimate and in the Soviet data. In four other cases there is perfect correspondence between our ordering and the indicators from the Soviet yearbook.

Consequently, one may conclude that the test of validation is passed quite successfully.

It is important to remember that the suggested ordering is made to indicate the burden of dated industrialization imposed on long-distance freight transport, mainly on railroads. From this point of view the arrangement of regions has a very logical structure. At the top of the list, with maximum burden, is the Donets-Dniepr region, the heartland of industrialization at the very beginning of this century, and later in the 1930s. Next comes Kazakhstan, scene of much forced labour from the 1930s to 1950s. It is a heartland of more recent efforts to replace worked-out mines of the Donets-Dniepr region. Vast investments came to Kazakhstan in the 1960s and 1970s in connection with the Soviet space programme, and even more – for the militarized complex which is safely far from international frontiers.

Freight carried by trucks in Central Asia and Transcaucasia is, obviously, very much orientated to the needs of rural places. The two regions are in a leading position in production of important crops (cotton, tea, grapes, etc.); they have dense networks of rural settlements and contrastingly low density of railroads.

Povolzh'ye, the Far East and the North-West also have a lot of dated industries. Characteristically the burden of outdated industries is felt much less in the Central region, where Moscow privileges permit the introduction of most of the innovations of the Information Age. Byelorussia and the Baltic region are also prominent in innovations, and it transpires in the positioning of those regions closer to the bottom of the list, at a low level of inconvenience from the Dated Legacy.

It is questionable that there are chances for Central Asia, Transcaucasia and Moldavia to bank their benefits of relative freedom from the dated legacy, and to absorb in a more rapid way the fruits of the Information Age. There are objective indications for developing versions of 'Silicon Valley' in those southern areas that have much surplus labour, but it's against the interest of political dominance of the Russian culture group in the Empire. On the other hand, it is mostly the Central Chernozem region which presently absorbs investments in computer production. It is also near the bottom of our list: it has a low level of interference from the dated legacy.

Our observations on the ordering of Soviet economic regions yield a number of valuable insights and more understanding of the structural difficulties in the Soviet economy. The observed phenomena are known as 'Regional pathology'.

Outdated activities in the USSR exhibit a dramatically significant pattern of location. The ordering of regions makes it clear that a corridor of difficulties has developed. It stretches from the Donets-Dniepr through Povolzh'ye and Kazakhstan to the Far East. This area from the Black Sea to the Pacific Ocean has, essentially, just one main rail line to serve it. It is impossible to find a worse example of the outdated legacy of industrialization in the territory of Russia. It shows no evidence of rational planning.

Appendix 2.1

List of Cities as Grouped in the $(x, y,)$ Space in Figure 2.2.

In the two-digit index assigned to each group, the first digit reflects the number of activities in a city with long-distance linkages, and the second digit reflects the preponderance of either secondary activities, 1 to 3, or tertiary activities, 5 to 8, with equilibrium designated by 4.

64 (one city)
Yakutsk

66 (one city)
Ashkhabad

53 (two cities)
Chimkent
Gur'ev

55 (four cities)
Alma-Ata
Batumi
Samarcand
Stavropol

48 (eight cities)
Chita
Dzhambul
Khabarovsk
Kustanay
Murmansk
Osh
Petropavlovsk/K.
Sochi

44 (nine cities)
Irkutsk
Kokand
Petrozavodsk
Simferopol
Sukhumi

Yuzhno-Sakhalinsk
Tselenograd
Ural'sk
Vologda

46 (five cities)
Dushanbe
Fergana
Ivano-Frankovsk
Kirovobad
Zhitomir

47 (one city)
Kirovograd

48 (one city)
Tbilisi

31 (two cities)
Pavlodar
Tyumen'

32 (one city)
Syktyvkar

33 (seven cities)
Arkhangel'sk
Bel'tsy
Brest
Khmelnitskiy
Noril'sk
Severodvinsk

Ussuriysk

35 (fourteen cities)
Andizhan
Astrakhan'
Blagoveshensk
Chernigov
Chernovtsy
Groznyy
Kutaisi
Kzyl-Orda
Leninabad
Makhachkala
Maykop
Namangan
Odessa
Rovno

37 (two cities)
Kishinev
Ordzhonikidze

22 (fifteen cities)
Berezniki
Bratsk
Kadiyevka
Kiselevsk
Klaipeda
Kommunarsk
Krasnyy Luch
Nakhodka
Novorossiysk
Orsk

Orsha
Serov
Sumgait
Volzhskiy
Yelets

24 (thirty-two cities)
Baranovichi
Belaya Tserkov
Berdyansk
Chirchik
Donetsk
Dzerzhinsk
Gor'kiy
Grodno
Kaliningrad
Karaganda
Kazan'
Kharkov
Kolomna
Kostroma
Krasnodar
Leninakan
Miass
Mogilev
Novorossiysk
Penza
Podol'sk
Poltava
Prokop'evsk
Saratov
Slavyansk
Tallin
Tambov
Ulan-Ude
Us't-Kamenogorsk
Vinnitsa
Vladivostok
Yoshkar-Ola

26 (ten cities)
Bukhara
Frunze
Kiev
L'vov
Lyubertsy
Moscow
Mytishchi
Novocherkassk
Vilnius
Yerevan

13 (sixty-six cities)
Aktyubinsk
Angarsk
Anzhero-Sudzhensk
Armavir
Balakovo
Barnaul
Belgorod
Belovo
Bryansk
Cheboksary
Cherepovets
Cherkassy
Daugavpils
Dneprodzerzhinsk
Elektrostal
Gorlovka
Ivanovo
Izhevsk
Kamensk-Ural'skiy
Kerch'
Kherson
Kirov
Kirovokan
Komsomol'sk
Konstantinovka
Kopeysk
Kovrov
Kramatorsk
Kremenchug
Kuybyshev
Leninsk-Kuznetskiy
Lipetsk
Lisichansk
Magnitogorsk
Makeyevka
Melitopol'
Nikopol'
Nizhniy Tagil
Noginsk
Novgorod
Novokuybyshevsk
Novoshakhtinsk
Orekhovo-Zuyevo
Orenburg
Perm'
Pervoural'sk
Petropavlovsk
(Kazakhstan)
Rostov-on-Don
Rubtsovsk
Rybinsk
Saransk

Semipalatinsk
Serpukhov
Shakhty
Sterlitamak
Syzran'
Taganrog
Temirtau
Togliatti
Tula
Ul'yanovsk
Voroshilovgrad
Yaroslavl'
Zaporozh'ye
Zhdanov
Zlatoust

15 (eighteen cities)
Biysk
Bobruysk
Engals
Kalinin
Kaliningrad
(Moscow Oblast')
Kaunas
Kemerovo
Leningrad
Minsk
Nal'chik
Pskov
Ryazan'
Sevastopol'
Smolensk
Sverdlovsk
Tiraspol'
Tomsk
Voronezh

04 (twenty cities)
Chelyabinsk
Dnepropetrovsk
Gomel'
Kaluga
Krasoyarsk
Krivoy Rog
Kurgan
Kursk
Nikolayev
Novokuznetsk
Novomoskovsk (Tula
Oblast')
Omsk
Orel

Riga   Ufa   Vladimir
Salavat   Vitebsk   Volgograd
Sumy

Chapter three

The interaction between industrial and social content of cities

Ideas leading to a study design

The functional typology of cities has performed its job in Chapter 2, and the next logical step is to shift our view to examine dimensions, that up to now have been left in darkness.

In Chapter 2, old functions of cities indicated by employment percentages were assessed, producing a portrait of Soviet cities according to perceptions of central economic planning. It is not the only important aspect of urban life. Chapter 1 disclosed influences dictated by a macro-structure of the urban network. However, up until now the human dimension of cities, their cultural manifestations, have not been discussed. It is time to fill this gap.

Chapter 3 attempts to highlight the human or social dimension of Soviet cities. We intend to assess the readiness of Soviet urban people for the post-industrial tendencies of the microchip age.

Soviet statistical data disclose very little about rank-and-file human actors in cities. It has sense to employ tools of statistical analysis in making conclusions about educational levels and urban cultural life. Human capital is of key importance in the Information Age, but it is a perishable substance; it needs daily cycles of regeneration. Human capital exists insofar as societies provide an interaction of talents, and it matures or degenerates according to the climate it faces or creates.

Leadership among innovative cities in the 1980s is no longer linked with sizeable employment. Automation and also large-scale readjustments of employment patterns have weakened the connection between city size and innovation. However, it is unclear whether post-industrial features are present in Soviet cities. Soviet urban employment has been shaped by a previous stage of history – by old-fashioned industrialization. Although the influence of industrialization on employment is clearly established (Harris, 1970), let us look at its dynamics with an open mind.

Our aim is to provide a study design that has a fair chance of detecting any strong post-industrial development in Soviet cities.

The study design adopted here is based on the assumption that not just industries are important in cities. Cities exist and grow partly as a result of economies of scale and from co-operation between industries. But they are in no way just centres of material production. Cities nowadays are also well-furnished centres of consumption in which social patterns are formed. They are also sites for the exchange of ideas, which makes them likely centres of innovation.

A political system's mechanism for demonstrating continuity is inspired by economics only in part; much of it is derived from those broader aspects of life which may be described as 'social'. Monuments commemorating historical milestones in the political development of nations are highly visible in major cities. In the Soviet Union such monuments are creations that have aggressively replaced the symbols of religions or of the Tzar's Empire: monuments to Lenin and the party and the KGB headquarters are interlaced with even more numerous monuments to industrialization.

The broader influence of cities on the life of the society has been highlighted in numerous studies in economics, sociology and geography in the latter case, mostly within schools of urban, social, and political geography (Berry and Horton, 1970; Rogers, 1971; Yu. Medvedkov, 1977, 1979, 1983; CNSS, 1980; Alaev, 1986). Geographers have the advantage of being able to relate different mechanisms or processes coexisting within a particular space (Berry, 1978; Lavrov *et al.*, 1979; O. Medvedkov, 1988).

Now we set about the task of relating two different profiles – industrial and social – within cities. Analysis of the interactions between these two profiles may lead eventually to better understanding of urban dynamics; at least we may recognize some of the social consequences of existing patterns of employment.

Input data

Our sample of 221 Soviet cities – all with 1970 populations of 100,000 or more – were used for this analysis as well. These highly developed, urbanized environments are in all parts of the country. The cut-off population size, 100,000, is high enough to guarantee that only truly urban centres, those distinctly different from rural settlements, would be considered.

The data form a 221×7 matrix – each of the 221 cities was provided with seven initial indicators. The inadequacy of published Soviet statistical data made it difficult to find even seven reliable indicators. Opportunities to evaluate the quality of the study data were afforded by the author's research position in the Soviet Academy of Sciences (1972–86). Consistency checks, discussions with experts and field observations were used to obtain better information than usually available.

Part of the data came from published records of the latest Soviet Census of Population (Chislennost 1984); the scarcity of these records is somewhat compensated for by accepting 'starting-point' data from more complete records of the previous census (*Itogi,* 1972; 1973; 1974). That census gave specific levels of education for residents in each Soviet city.

Numerous directories listing educational and other cultural facilities in Soviet cities form another category of data sources. They make possible the calculation of an aggregate measure for a 'cultural potential' index for each city – a CPI. The CPI was initially devised and computed by Stella Axelrod (1978) (presently, Mrs Kreinin, Rehovot, Israel) during her work in the Moscow Central Institute of Urban Planning. CPI values are calculated by an algorithm which clusters cities in n-dimensional space of culture-related urban features (expressed in a quantitative or qualitative way). The systems of measurement used for scaling the values were selected to maximize the differentiation among cities, a practice usual in the classification techniques (James, 1985).

The study design arranges the seven items of data assembled for 221 cities into two distinctly different sets. Four of the data items are percentages of employment. That reflect basic urban activities:

M – manufacturing
R – railway and other transportation
S – scientific research, pure and applied (R&D)
A – administration

The industrial profiles of Soviet cities as represented by the relative size of these four aggregated employment sectors are informative. For example, they make it easy to identify widely known imbalances in major functions among Soviet economic regions (Shabad, 1985; Dienes, 1987; Panel on the Soviet Union, 1987).

The use of employment statistics (M, R, S, A) in this study clears the way for examining the legacy of the industrialization in Soviet cities. However, this category of data is not the only one used in the analysis; the objective, in fact, is to counterpoise it with another data set.

Three other indices, sociocultural characteristics of Soviet cities, act as the counter-weights. These indices, designed to reflect cities' cultural, intellectual, or leadership achievements, should make it possible to detect any post-industrial features present. Specifically, the indices are:

U – percentage of the labour force with a university level education;

C – the level of cultural potential, by CPI (which, as explained above, measures the diversity of educational and other cultural institutions in each city); and

T – the growth rate of the population between 1970 and 1979, (a ratio of population increase to initial population).

The last index, T, indicates in a general way the presence of all sorts of reasons for growth in cities. Because the period under study was one in which Soviet industrialization slowed markedly (1970–9), high values of T point to growth in a post-industrial phase.

To reduce the complexity of interactions among the data selected, a time lag was introduced into the group of indices. Influences travel in time only from the past to the present. The effect of influences from the past is taken into consideration through the use of more recent values of C and T than for M, S, A, and U. These steps were taken to enhance the following one-way relationship:

$$(M,R,S,A) \rightarrow (U,C,T) \tag{3.1}$$

e.g., the effect of employment structure on the sociocultural characteristics of the cities. Expression (3.1), however, is nothing more than an expression of the direction of influences. A method is needed to describe these influences in an objective and exact way.

Method of correlating one data set with another

All multivariate classifications are based on data compression techniques. In urban analysis numerous features of each city are converted into co-ordinate values that describe a city's location within some artificially created multidimensional feature of classification space. Systems of measurement in this space are defined by the co-ordinate axes, which are selected and assigned scales in such a way as to encompass the maximum amount of variance in the initial data. Thus the method of data measurement is a crucial aspect of the study.

In this study the method is practically predetermined by the study design. To examine how one set of data projects into another, one may use as identical formulation from the classical statistical essays, particularly the Hotelling (1936) essay on canonical correlation analysis. This same method was used in Chapter 1.

Hotelling's analysis may be viewed as a technique for developing a set of optimally informative indices. The most innovative yet simple idea may provide the best indices, as evidenced by the straightforward, simple method of constructing 'weighted indices' in econometrics. These are the indices used in the daily press to track economic performance. In respect of the two data sets (taken separately) in the present study, formulae (3.2) and (3.3) describe a generally-accepted method for merging the data:

$$(w_1M + w_2R + w_3S + w_4A) = Z; \ (W_1U + W_2C + W_3T) = X \tag{3.2}$$

Hotelling's procedure permits an objective and optimal selection of the weights, w_i and W_j ($i = 1, \ldots, 4; \ j = 1, 2, 3$). Hotelling suggested calculating weight coefficients (w_i, W_j) by maximizing dependence between the two data sets (in our case, between M, R, S, A, on one hand, and U, C, T on the other). In other words, the resulting values Z and X are designed to tell the influence of one data set on another.

Details of the procedure underlying the analyses are illustrated in Table 3.1.

The seven initial variables comprise seven columns, each consisting of 221 lines (representing the 221 cities). Similarities between each pair of the seven columns are examined initially. Simple linear correlation coefficients are computed for this purpose, generating a (7 × 7) matrix of intercorrelations. Table 3.1 presents the (7 × 7) matrix R as a structure of four blocks:

$$\begin{bmatrix} R_{11} & R_{12} \\ R_{21} & R_{22} \end{bmatrix} \tag{3.3}$$

Table 3.1 Correlations among initial characteristics of cities

Variables	A	S	R	M	U	C	T
A	1	0.116	0.191	−0.625	0.295	0.173	0.282
S	0.116	1	−0.162	−0.204	0.239	0.490	−0.165
R	0.191	−0.162	1	−0.583	−0.58	−0.058	0.21
M	−0.625	−0.204	−0.583	1	−0.215	−0.161	−0.367
U	0.295	0.239	−0.158	−0.215	1	0.258	0.243
C	0.173	0.490	−0.058	−0.161	0.258	1	−0.132
T	0.282	−0.65	0.121	−0.367	0.243	−0.132	1

It is convenient to discuss the subsequent computations and the logic thereof as a succession of steps, as follows:

1. The creation of four blocks, denoted by R with subscripts, is justified by the objective of studying the relatedness of the two profiles in cities. The 'northwestern' block, R_{11}, indicates the degree of similarity among cities on the initial measures of the industrial profile. From this block it is possible to determine whether or not the profile is indistinct and/or multifaceted. The same information about the social profile is reported by the 'south-eastern' block of Table 3.1, that is, by values for R_{22}. The influence of one profile on another are revealed by the two other structural blocks of Table 3.1, those which do not belong to the main diagonal of (3.3). The fact that $R_{12} = R_{21}$ illustrates that the correlation coefficients cannot be used to infer the direction of influences: they measure the presence of influences only.

2. Then eigen values $(X*_1)$ and eigen vectors $(\mathbf{V}_1)$ of matrix A, $A = (R_{22})^{-1} R_{21} (R_{11})^{-1} R_{12}$, are calculated, assuming $X*_1 =$ max.

3. The first eigenvector $\mathbf{V}_1$ (with its $\mathbf{V}_1$ and $X*_1$, where $\mathbf{V}'_1$ is transposed $\mathbf{V}_1$) is used to obtain weight coefficients in the form of vector-columns:

$$W = \mathbf{V}_1 (\mathbf{V}'_1 R_{22} \mathbf{V}_1)^{-0.5}, \ W = (R_{11}^{-1} R_{12} W_1)/(X*_1)^{0.5}.$$

4. Then the main results – the canonical indices Z and X are calculated using the formula $Z = Q_1 w_1$ and $X = QW_1$, where Q_1 denotes the statistically standardized initial data in the form of 221 × 4 matrix with columns (A, S, R, M); statistically standardized initial data for the columns (U, C, T) which form a 221 × 3 matrix, are denoted by Q.

5. In order to simplify the interpretation of indices Z, X, the factor structure is determined by specifying correlations between indices Z and X and initial variables $S1 = (R_{11} \ w_1)$ and $S2 = (R_{22} \ W_1)$.

6. The informational value of the indices – that is, their ability to highlight a variety of cities – is then analysed. The percentage of variance 'accounted for' in the initial data is calculated by the formula:

$$P1 = (100/n) \ S1^2_i; \ P2 = (100/m) \ S2^2_j; \ (n = 4, m = 3).$$

7. If the percentage of variance is not sufficiently high, i.e., not significant, procedures (3) to (6) are repeated for the next eigenvalue of A with its eigenvector.

Table 3.2 Factor structure and weight coefficients·

| Variables | Correlation between indices and variables | | | | Weight coefficients for the indices | | | |
| | Development | | Growth | | Development | | Growth | |
	Z1	X1	Z2	X2	Z1	X1	Z2	X2
A	0.618	—	−0.351	—	0.315	—	0.185	—
S	0.796	—	0.541	—	0.606	—	0.896*	—
R	−0.109	—	−0.189	—	−0.350	—	0.751*	—
M	−0.595	—	0.517	—	−0.479	—	0.897	—
U	—	0.664	—	−0.090	—	0.394	—	0.120
C	—	0.826*	—	0.453	—	0.735	—	0.420
T	—	0.401	—	−0.905*	—	0.328	—	0.897

Table 3.2 presents values of correlation coefficients for the initial data and the canonical indices.

The left part of the Table 3.2 tells how strongly the canonical indices correlate with the initial variables. It provides a profile for each canonical index, in much the same way as happens with factor loadings in factor analysis. If the reader is familiar with factor analysis, no further explanation of the left side of Table 3.2 is necessary.

For other readers, a more detailed explanation follows. A number of values on the left side of Table 3.2 are close to either +1.000 or −1.000. These are cases of almost perfect projection of initial data into the canonical indices. The indices, similar to factors or components, are designed to highlight a simple structure, which might exist, in latent form, in the array of initial data. In order to ascertain what this structure is, it is necessary to determine which initial variables project into which indices. High correlations between initial variables and canonical indices send messages about the meaning of one or other index, i.e., they are signals about the relatedness between an index and the correlated variable, which together with other variables correlating closely with the index, may suggest a name for the index. For example, growth rates in city sizes (T) have a tight correlation (−0.905) with the canonical index X2. This particular index has no other remarkably strong correlations. Thus, in interpreting the meaning of X2 one must be guided mostly by its association with T. This is an indication that X2, as an index, reflects mostly the mechanism of urban growth.

Two mechanisms influencing dynamics of the cities

The main conclusions reached by means of canonical correlation analysis indicate that the economic and social characteristics of Soviet cities influence different aspects of urban change. Two directions of influence emerge as soon as one starts computing the explained canonical correlation indices from the data set: (a) a mechanism for socio-cultural prominence of a city; and (b) an impact of employment profile on changing population size.

The linkages between the two aggregated indices, Z and X, do not provide an adequate description, nor can Z $\rightarrow$ X; in order to embrace a major part of variance in initial data, one must compute two sets of indices. In other words, Z must be split into Z1 and Z2, and X into X1 and X2. This differentiation makes it possible to highlight two paths or directions of influence, with two separate relationships for the indices:

Z1 $\rightarrow$ X1
Z2 $\rightarrow$ X2

The first of the equations is interpreted here as a mechanism of modern development. It incorporates 30 per cent from the initial data set. The second equation, according to my interpretation, describes a mechanism of industry related growth. Its share of the variance of initial data is 25 per cent. The two mechanisms manifest themselves with approximately the same strength: that is the main message. The near-equality of the two mechanisms means that our initial and intuitive dissatisfaction with purely economic functions of cities as the basis for functional classifications is well supported by data.

Both sets of canonical indices, (Z1, X1) and (Z2, X2), and what they say about urban dynamics, may be computed for each city. These measures allow us to analyse the distribution of all 221 cities in a co-ordinate space so that we may observe the cities within type-defining frameworks (Figure 3.1). The co-ordinate spaces of Figure 3.1 are determined by axes which correspond to the paired indices: (Z1, X1) and (Z2, X2). Each time the axes are calibrated in units equal to a standard deviation for specific city values $Z1(i)$ and $X1(i)$ or $Z2(i)$ and $X2(i)$, where $i = 1, 2, \ldots, 221$. The values projected on the axes thus belong to cities, and they are statistically normalized (for each axis the average equals zero).

This process provides a graph in which each city appears in the form of a point. If a point falls to the right of 0 on the Z1-scale it signifies that it has above average development compared with the entire set of cities. If a point representing a city falls above 0 on the

a

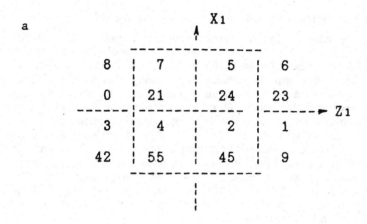

b

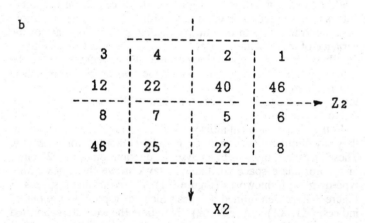

Figure 3.1 Distribution of cities in co-ordinated spaces defined by the canonical indices (a) Z1, X1 and (b) Z2, X2.

L₁

Number of cities

	46	40	12	22	22	8	25	46	Number of cities
7	2	4	0	0	2	2	1	10	21
6	6	3	2	2	1	1	2	0	25
5	4	5	0	0	3	1	1	10	24
4	5	12	2	8	9	1	9	9	55
3	22	9	2	3	4	1	0	1	42
2	3	7	6	7	3	2	11	6	45
1	4	0	0	2	0	0	1	2	9
	1	2	3	4	5	6	7	8	L₂

(left axis: Segments on the L1-scale)

Segments on the L2-scale

Figure 3.2 Distribution of cities based on two-digit codes derived from their position in the co-ordinated spaces of Figure 3.1.

Z2-scale, its mechanism of industry-related growth is below average (Figure 3.1).

Sub-fields of the graphs shown in Figure 3.1 may be numbered, providing each city with a two-digit code. It makes it possible to construct another graph (Figure 3.2). The first digit, on the L1-scale, designates the impact of the economic profile on the modern development of cities; the second digit, on the L2-scale, designates the impact of economic activity on the city's growth.

Movement to the right on the L2-scale corresponds with the increasing influence of economic life on city growth; movement up the L1 scale corresponds with the increased influence of economic activity on modern development among cities. In other words, the cities in more favourable situations will be found in the upper part of this table, and a few – in the most favourable position – in the north-eastern corner. Four very strongly divided clusters are identified using the numerical taxonomy algorithm, equivalent to sorting all of the studied Soviet cities among four boxes. Appendix 3.1 lists each city according to its affiliation with one of the following types:

1. Type P_{01} – restricted growth with significant development (twenty-eight cities, including Moscow, Kiev, Tbilisi).
2. Type P_{11} – significant, both growth and development (forty-two cities, including Minsk, Samarkand, Fergana).

3. Type P_{00} – restricted, both growth and development (ninety-two cities, including Tula, Volgograd, Donetsk).
   Characteristically this patently stagnant group represents main centres with specialized profiles shaped by the Soviet industrialization programmes.
4. Type P_{10} – significant growth with restricted development (forty-nine cities, including Nakhodka, Cherepovets, Kemerovo and all the rest of the cities shaped largely by Soviet industrialization).

One might conclude from these findings that the legacy of industrialization already serves the studied Soviet cities poorly. They look as if they are already in a phase of post-industrial readjustment. There may be, however, an explanation of another sort: the unfavourable dynamics of Soviet industrial centres may originate instead in structural deficiencies within Soviet society, which has managed its industry poorly during a time of depletion of both labour and mineral resources. Before we seek the origins of the four types of Soviet urban dynamics, however, we need to validate our categorization of types of dynamics.

Scepticism is essential to science. With this in mind, let us assume a 'devil's advocate' position, which may be summarized as follows: 'These four types have been derived from data that are sketchy and outdated; what relevance or meaning do they have today?'

Our response: their validity can be tested. Such a test is summarized in Table 3.3. Recent data on population trends of Soviet cities (Shabad 1985) are employed, as well as data on population growth in 1986, in order to check the extent to which the actual performance of Soviet cities during the period 1979 to 1984 and 1979 to 1986 correspond to the performances indicated for the four types of dynamics.

The correspondence is perfectly satisfactory. According to our classification, the two types of dynamics associated with lack of

Table 3.3 Verification of urban dynamics

Types of urban dynamics	Median growth rates 1979-84 %	1979-86 %
P_{00}	5	7
P_{01}	7	9
P_{10}	10	14
P_{11}	14	20

growth (P_{00}, P_{01}) demonstrate the lowest median growth rates in the 1980s. The other two types show median growth rates two times larger, and their ordering by growth rates is also exactly the one which the logic of our types predicts: low presence of development in P_{10} leads to 10 per cent growth between 1979 and 1984 and to 14 per cent between 1979 and 1986 in city sizes, as compared with 14 per cent (1979–84) and 20 per cent (1979–86) for the most favourable situation, P_{11}. The categories of urban dynamics thus identified are obviously robust and enduring.

Another perspective on the problem

The next step of the study entails taking a different perspective. The overall study design boils down to a search for the general factors responsible for differential rates of growth and change in Soviet cities. Yet such a study need not and should not be restricted solely to an investigation of features of industrialization time or post-industrial characteristics. Additional influences which may create problems for Soviet industry should also be considered. Within the economic and social environment of the USSR are many threats to cities shaped by the industrialization policies of the past. This legacy is currently the subject of much debate in the Soviet press, on pages set aside for the official policy of controlled 'glasnost'. Generally, what one reads is a litany of Soviet structural inefficiencies.

Geographers may pinpoint some of these structural inefficiencies by applying instruments of either regional or spatial analysis. Here we will focus first on tools of regional analysis, specifically, on aggregating facts according to the network of major Soviet economic regions; this focus permits us to analyse territorial structures larger than the individual cities. Next we will identify those territorial structures which have been generated or radically modified by the Soviet system of centralized planning. Our third step is to look at the influence of these structures on the four types of dynamics in Soviet cities (P_{01}, P_{00}, P_{10}, P_{11}).

The author's prior understanding of the Soviet situation, suggests another category of territorial structures relevant for this analysis – those suspected of being mismanaged or otherwise disrupted but which still have potential to influence Soviet urban life.

Territorial structures which affect cities

Both management of resources and organizational settings of Soviet industrial centres follow distinct geographical patterns. On the one hand are patterns generated by nature, which dictates that many centres be located in areas rich with resources. On the other hand are forces and locational factors generated by industries themselves. The question thus arises, what kind of territorial structures should be selected when one is seeking numerically precise characteristics based on the statistics available and at the same time trying to determine their potential impact on the performance of cities?

The following structures will be discussed here:

G1 – clustering (to describe urban networks in a region);
G2 – rural vitality (to describe rural areas around cities);
G3 – connectedness (to describe opportunities for cities to interact with each other); and
G4 – marginality (to assess the macro-location of cities in terms of centre/periphery contrasts).

Numerically precise indices or indicators are necessary for each of the structures, in each case, for every major economic region of the Soviet Union. In the years, which are covered by classification of cities (1970 to 1983), there were nineteen such regions (a twentieth economic region, the north, was created from a portion of the north-west economic region in 1983).

In selecting the indicators, we remember the necessity of allowing a time-lag between measurements of the phenomena that we consider influences and the indices actually representing the results of those influences. The one-way influence of the past on more recent events is what must be measured. For this reason all indicators used to describe G1–G4, were derived from data of the 1960s. From that not very distant past, a period filled with major Soviet construction projects, one may expect to find influences on our types of urban dynamics ($P_{01}, \ldots, P_{10}$) that have been defined for the 1970s and 1980s.

The following criteria were employed in constructing indices for G1–G4:

1. No more than three indicators were included for each structure. This limitation comes from the regression techniques employed, which do not accommodate more than three indicators very effectively when the number of regions is small as 19.
2. The indicators were aggregated in time into a single numerical

index, one for each specific territorial structure, in a
framework of a linear regression model.

3. Indicators for the indices were selected through trial and error
   so that they demonstrate acceptable success of predicting the
   regional distribution for at least one type of cities (P_{01}, . . .,
   P_{10}).

In other words, multiple regression analysis was used for clustering
some initial indicators into an index. As a result, four columns of
indices are generated for G1–G2.

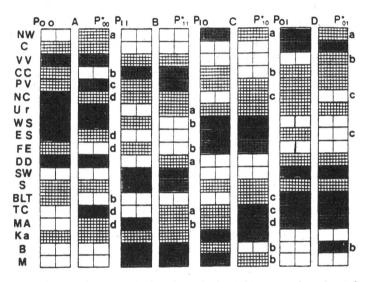

Figure 3.3 Four types of urban dynamics in major economic regions of
the Soviet Union. The four successful tests are displayed as four pairs of
bar graphs: A, B, C, and D. The left bar graph in each pair reflects the
actual regional distribution of cities with a particular dynamics: P_{00}, P_{11},
P_{10}, P_{01}. The density of the patterns within the bar graphs corresponds to
percentages of cities with one or another type of dynamics in nineteen
regions (dense pattern = high percentage, sparse pattern = low percentage). The right bar graph in each pair (the one marked by a star) shows
percentages predicted by the multiple regression analysis ($P*_{00}$, $P*_{11}$, $P*_{10}$,
$P*_{01}$). Correspondence between the actual and predicted percentages is
very good. A key to abbreviations to the economic regions' names appears
in Appendix 3.1. Lower case letters to the right of each bar graph identify
a source of interference from minor factors: (a) widely scattered urban
agglomerations; (b) compactness of the region; (c) magnitude of the
region; and (d) ribbons of cities.

Numerical values of the indices are designated here as $P*_{00}$, $P*_{01}$, $P*_{11}$, $P*_{10}$, in order to emphasize the fact that these numerical values are less general than the initial structures, G1–G2. Regression analysis acts here as a filter: it eliminates information about the structures G1–G4. These structures are reproducing the regional presence of one or another type of city (P_{00}, P_{01}, P_{10}, P_{11}).

A classic situation of hypothesis-testing is evident here; the indices pass through a filter of four tests as a check on their validity. For example, in a G1 test, each of the types of urban dynamics is checked to see how well its presence in economic regions may be predicted by observing the clustering levels of urban networks in each region. Each test (G1–G4) demonstrates how well the indicators can predict the regional distributions of the four types of urban dynamics.

In this light the discussion which follows may be shaped according to a framework of hypothesis testing. This testing uses an analysis of variance (AOV) attached to standard programs of multiple linear regression. Each time 'acceptable success' is defined as a more than 0.95 probability that linear dependence exists between the indices ($P*_{00}$, ..., P_{10}) and the real regional percentage of cities with one or another type of dynamics.

Figure 3.3 summarize the results of the tests with the help of bar graphs.

Influences of clustering on urban dynamics

The phenomenon of clustering (G1) was examined using following indicators:

$N*$ – number of cities with 1970 population above 100,000 in economic regions.

A_c – Ekkel's coefficient of clustering (Ekkel, 1978). Ekkel's indicator is constructed by calculating percentages of total overlapping zones of influence of settlements over 15,000 population.

k – Polyan's coefficient of complexity for urban agglomerations (Lappo, 1978), which combines percentages of population in the core city and its satellites. It is somewhat modified insofar as it is standardized according to the number of large cities in each region.

The formula for computing $P*_{00}$, which yields a figure very close to the actual percentage of P_{00} in every economic region, is:

$$P*_{00} = 10.13835 - 1.69802(k) + 45.6766(A_c)$$
$$- 0.12686(N*) \qquad (3.4)$$

(Fisher coefficients specific to each predictor of clustering are: $F(k) = 0.222$ $F(A_c) = 14.460$ $F(N*) = 1.156$; for all three predictors in their interaction, $F = 4.899$.)

This result means that restricted growth combined with restricted development, a feature of many Soviet cities is explained by something besides the post-industrial phase. The statistical quality of formula (3.4) is quite good: it is characterized by Fisher criterion, $F = 4.899$, which is well above the tabulated critical level $F_{(tab.95)} = 3.290$.

As the first two bar graphs (Figure 3.3) demonstrate, there is a strong similarity of the columns, $P*_{00}$ and P_{00} (the predicted and real-life percentages of this particular unfavourable type of urban dynamics).

The only major discrepancies are presented by cities in Transcaucasia (Tc), where mountains make it difficult for centralized planners to dictate the clustering of cities.

Another notable aspect of (3.4) is its coefficient of determination: $R^2 = 0.49$. In other words, almost half the variance for regions in frequencies of cities of type P_{00} may be explained by levels of clustering. The ability to explain that much cannot be disregarded.

The misfortune of type P_{00} is widespread; these conditions affect ninety-two cities, or 42 per cent of all cities which, as early as 1970, had population above 100,000.

The most spectacular result, however, is the fact that the statistical tests have proved that clustering exerts a very selective influence; its response, $P*_{00}$, appears only in cities with type P_{00} dynamics. We could not detect statistically significant linkages between clustering and the other three types of urban dynamics.

It is rare that a mechanism of influence is isolated so clearly, *one* factor having an effect on *just one* type of urban dynamics. The origin of a particular urban dynamic, P_{00}, can be traced to distinct geographical structures present as early as the 1960s and it is not found in any regions with the features of the post-industrial phase.

We deal here with the misfortune of numerous relics of Soviet industrialization. Our analysis tells us that these cities suffer from chaotic sprawl, and from conflicts or complexities in land-use, specifically in zones where the built-up space of neighbouring settlements is merging (as is described by Ekkel's indicator).

Among the predictors present in (3.4) the biggest influence comes from A_c, a measure for urban agglomerations. The A_c –

specific value of F is substantial: it indicates that most of the success of (3.4) may be attributed to A_c. The same picture is provided by the corresponding partial correlation coefficients:

$$R\ (k, P_{00})\quad = 0.0865$$
$$R\ (A_c, P_{00})\ = 0.6000$$
$$R\ (N*, P_{00})\ = 0.3290$$
$$R\ (k, A_c)\quad = 0.6212$$
$$R\ (A_c, N*)\quad = 0.7002$$

It is obvious that A_c encompasses influences which may travel from k and $N*$ toward P_{00}.

It is remarkable that Ekkel, who developed this powerful indicator of clustering, considers it – with no knowledge of our study – to be a right measure for complexity and conflicts of land use in merging settlements (Ekkel, 1978).

Soviet geographers frequently debate positive and negative aspects of the increasingly obvious existence of urban agglomerations. Our results indicate that such agglomerations have a substantial negative influence.

Anything more positive than the P_{00} dynamic is not influenced by the clustering phenomenon:

$$P_{01} = f(K, A_c, N*)\quad F = 0.508;$$
$$P_{10} = f(K, A_c, N*)\quad F = 0.763;$$
$$P_{11} = f(K, A_c, N*)\quad F = 1.534;$$

Influences from the other three territorial structures

The second territorial structure, that of rural vitality, (G2), is studied here in connection with six indicators:

V – rural population density, 1966 (residents/km^2)
Z – rural population density, 1976 (residents/km^2)
C – change in rural population density, 1966-76 (%)
S – cultivated land, 1966 (millions/ha)
L – rural population per unit of cultivated land in 1966 (residents/ha)
I – labour-intensity in agriculture, by dummy variables (1 – intensive surburban, 2 – multi-profile, 3 – monoculture)

An effort was made to choose the kind of indicators that would reflect the ability of rural areas to provide the flood of labourers into the cities to increase their populations. Calculation of the first five indicators from 1966 to 1979 takes into account the lag in time; it affords an opportunity to define the influence of rural population structures on immigration processes.

As to the value of I, this estimate was obtained by expert evaluations; it helps to understand the idea of specialization in rural areas in a very vivid way.

The calculations show that the phenomenon of rural vitality is definitely linked with the evolution of cities in the case of P_{11} cities – those with significant development and growth. The evolution of the other types of settlement are not predetermined by the phenomenon of rural vitality. Four regression equations were calculated; one of them gave the best results:

$$P_{11} = -25.431 + 14.395(I) + 0.856(V) - 0.024(C);$$

$$F(I) = 3.029; \ F(V) = 17.891; \ F(C) = 0.008;$$

$$R^2 = 0.580; \ F = 6.976 > F_{tab} = 3.290 \qquad (3.5)$$

This formula explains more than half of variance (58 per cent) in the percentages of P_{11} cities registered in nineteen major economic regions. The F-criterion is more than twice the value in the table. Another feature of this equation – low correlations among I, V, and C, which allow us to avoid the effect of multicollinearity.

The evolution of cities with significant growth and development (P_{11}) is correlated most of all with I – the type of agriculture in the region and with V and Z – density of rural population. It's not a surprise that cities of this type are widespread on the south and south-west of the Soviet Union, where rural density is very high and agriculture is much more developed than in the other regions.

The graphs in Figure 3.3 register the similarity of patterns for the presence of rural vitality features and for P_{11}. Urban dynamics of the most active type, P_{11}, are present mostly in Byelorussia, the south-west, Moldavia and the south, as well as in central Chernozyom.

These are the regions which in the 1970s still could provide sizeable migrations from rural areas to local cities. Here is an example of the most basic and simple dependence on specific features of rural areas. The rural population, tired of the rigid regimentation of the *kolkhozes* (centrally planned collective farms), has continued its exodux, hoping to find something more attractive in the cities. That is the essential influence on the dynamics of the type P_{11}. There is little here to diagnose as clear-cut traces of the post-industrial phase.

The formula above accounts for 58 per cent of variance in indicators of the presence of urban dynamics of type P_{11} in Soviet economic regions. Thus the bulk of the variance is accounted for

109

by a single territorial characteristic of rural vitality. It's a very graphic result.

The third territorial structure to be tested is G3, the 'connectedness' of each region. This term refers to administrative provisions created by the Soviets to enhance co-operation of cities within each region's boundaries. Indicators tested are:

N – number of capitals of union republics in the economic region (such centres have bureaucracies with the authority to establish intraregional co-operation)

M – number of radiating railroads, average for all local groups of cities in the region (Davidovich, 1976)

B – number of neighbouring administrative-territorial units – one step below the economic regions in the political-administrative hierarchy – the so-called beta-index of the graph theory (Haggett and Chorley, 1969; Haggett *et al.*, 1977) is used here:

$I*$ – an index of the shape of a region's territory (a measure of how far it diverges from the maximum compactness of a circle):

T – number of shared railway segments in all local groups of cities in a region (Davidovich, 1976):

R – number of railways crossing the region's boundary.

The first two indicators (N, M) turned out to be of the most important of all tested regression formulae. To provide the lag in time, the values for M and I reflect in each case the situation of the end of the 1960s.

Our examination of regression formulae for 'connectedness' has produced a very clear message. As with our territorial structures, 'connectedness' generates just one type of urban dynamics. The type in this case is P_{01} – restricted urban growth with the prominence of the sociocultural performance. P_{01} is the very type of dynamics which might be attributed to the post-industrial phase of a society; but such a conclusion would be incorrect in the present situation, because the factors responsible for P_{01} lie in structures of quite another sort. Influences from 'connectedness' are the most likely explanation. The influences correspond to the formula:

$$P_{01} = -6.363 + 0.370(N) + 9.350(M) \qquad (3.6)$$

$$[R^2 = 0.43; F = 6.042 > F_{tab} = 3.630$$

$$F(N) = 7.150; F(M) = 4.933]$$

Inasmuch as F(N) is large, transport infrastructure (railways), appears to play a key role in the co-operation of settlements on an intraregional level. It is clear that regions which are rich in this infrastructure also have the P_{01} type of urban dynamics.

Bar graphs and maps (Figure 3.3 and 3.4–3.7) demonstrate that actual frequencies of cities with sociocultural prominence are well-predicted by connectedness. In particular, the correspondence is perfect in regions such as the central, the north-west, and the south-west, in the Baltic Republics, in Transcaucasia, and in Middle Asia.

The fourth and the last territorial structure to be tested in G4, the presence of 'marginality' in the macrolocations of cities in one or another of the economic regions. In essence, it is an inquiry on how far the region's main centre is from the core of the nation or from lines of contact with other cultures. The following indicators tell the story of 'marginality':

C – expenses of rapid travel between Moscow and the main city in the region (price of regular air-tickets, in rubles):

$C*$ – the same travel, measured in km, by railway:

$T*$ – the same air travel, measured in hours of flight:

H – proximity to the contact zone with southern Islamic nations (degrees of latitude are used to represent the distance):

E – distance from the contact zone with European culture (degrees of longitude are used in this case):

Conventions accepted to measure H and E are set in a manner that provides an increase in 'marginality' in the eastward and south-ward directions. In this way these two indicators of 'marginality' reflect opportunities for involvement of an economic region with the European cultural heritage. In Russia such chances usually diminish as soon as one moves to the east from Poland or to the south from the Baltic sea-coast.

The best formula which was computed demonstrates strong connections between the indicators of marginality and one particular type of urban dynamics: P_{10}. In other words, marginality is associated with significant growth and restricted sociocultural performance of cities. The formula is as follows:

$$P_{10} = 12.123 + 0.584C - 5.574H \qquad (3.7)$$

$[R^2 = 0.523; F = 8.776 > F_{tab} = 3.680$

$F_c = 1.570; F_H = 0.452]$

The P_{10} type of dynamics primarily affects newly formed cities located in frontiers far from historical European cultural core.

111

GROWTH : NO DEVELOPMENT : NO

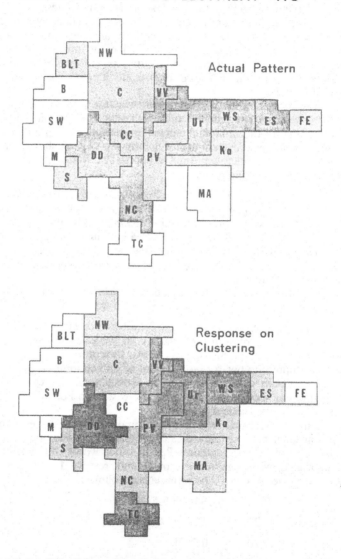

Figures 3.4–3.7 Schematic maps of urban dynamics in major economic regions of the Soviet Union. The size of each region on the map is proportional to its 1987 population.

GROWTH:NO   DEVELOPMENT:YES

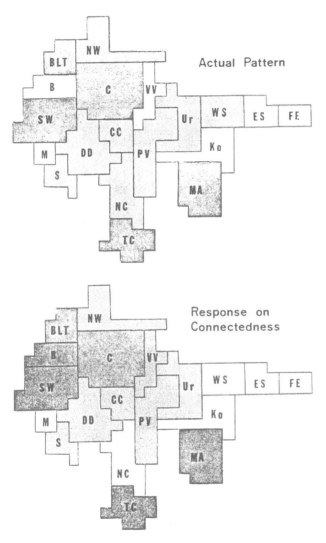

Actual Pattern

Response on
Connectedness

GROWTH : YES   DEVELOPMENT : NO

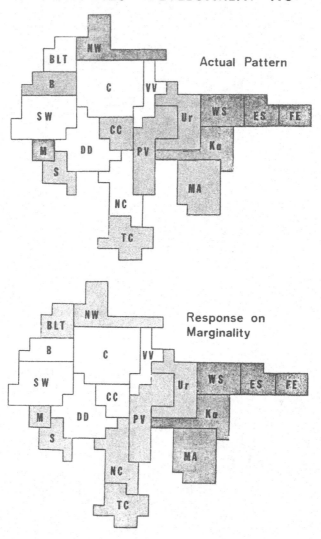

Actual Pattern

Response on
Marginality

GROWTH : YES   DEVELOPMENT : YES

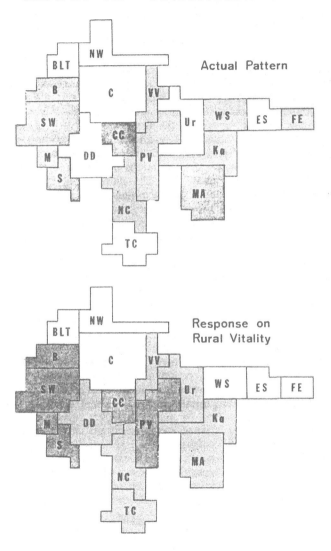

Actual Pattern

Response on
Rural Vitality

Conclusions

A review of our findings shows that four types of urban dynamics were identified, designated by the symbols P_{00}, P_{01}, P_{10}, P_{11}. They indicate, first, that cities perform differently in terms of growth and development. Second, they reveal the existence of an awkward situation, one of only slight growth and sociocultural development, which occurs mostly in centres with narrowly specialized employment profiles. Finally, we find that non-industrial cities, which have mainly sociocultural functions, perform better.

As one way of explaining the three findings, i.e., of relating them to one another, one might look at the general entrance of the Soviet Union into post-industrial life. Although the author's opinion about that is negative, opinions must be strengthened by facts; so more facts about the 221 cities studied here were examined to shed more light on the origins of the dynamics.

If a number of cities show evidence of post-industrial change, if they develop mostly as a result of their socio-cultural prominence, does it also mean that all other cities are substantially affected by the demise of the industrialization phase?

When the question is formulated in such a general form (made deliberately too general), then it is obvious that a fast answer is too risky. First of all, it has been determined with certainty that different types of Soviet cities are experiencing different fortunes at the present time. What generates an unfortunate situation for industrial cities is not necessarily the arrival of the post-industrial phase. There is an equal or greater chance that the cause may be inability of the Soviets to run their industrial empire properly; an accumulation of bad decisions may have shaped that empire.

Exactly that kind of answer came to light during the assessment of influences from territorial structures present in major Soviet economic regions.

The main message from our analysis is that it is too early to apply a postindustrial explanation for the misfortune of Soviet cities which have little or no development and growth. There are geographical structures created by society which can explain much of the difficulty. Such structures may be described as four different networks of cities shown on Figure 3.8–3.11:

(a) ninety-two of restricted growth and development;
(b) twenty-eight cases of restricted growth and significant development;
(c) fifty-nine cases of significant growth and restricted development; and
(d) forty-two cases of significant growth and development.

In a detailed and specific way it was possible to attribute the unlucky dynamics to the pattern of Soviet economic regions, a pattern which more simply (and less accurately) may be described in the following statement: labour force reserves, fixed capital, and raw material are all separated in different corners of the empire. As the description of the 'corridor of difficulties' at the end of Chapter 2 suggests, it's hard to bridge together the macro-regions of the Soviet Union.

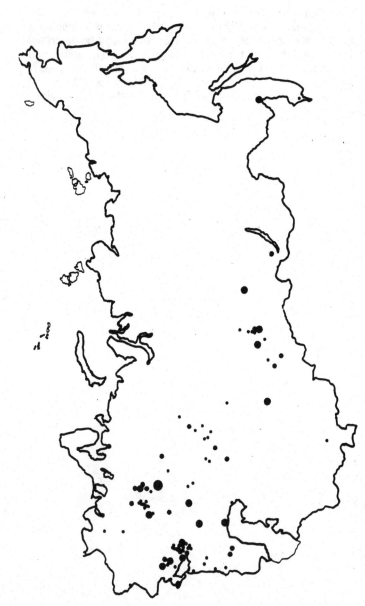

Figure 3.8   Cities with restricted growth and development, P$_{00}$-

Figure 3.9 Cities with restricted growth and significant development, P_{01}.

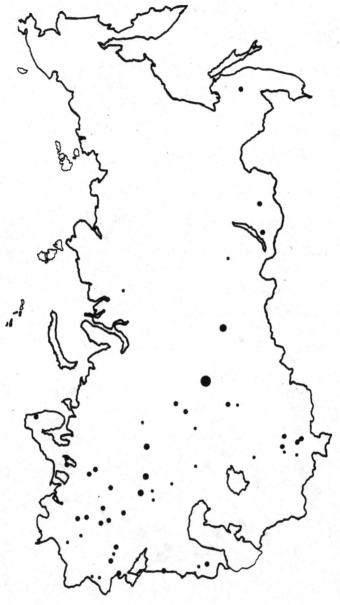

Figure 3.10   Cities with significant growth and restricted development, P_{10}

Figure 3.11 Cities with significant growth and development, P_{11}.

Appendix 3.1

Types of urban dynamics in the Soviet Union

The table specifies four types of dynamics for 221 cities in the USSR, all with populations above 100,000, beginning in 1970. The types are explained in the main text. The order of cities within each section in the table is based on scanning the national territory from north to south, starting from the west. The first entries in each section belong to the Russian Federated Repulic (RSFSR); smaller Union Republics follow. Within each region cities are ranked according to their sizes in 1984. Location of cities is specified up to the level of main economic region. For this purpose we use the following abbreviations:

Abbreviation		Region
	RSFSR	
NW		North west
C		Central
VV		Volgo-Vyatka
CC		Central Chernozem
PV		Povolzhye
NC		North Caucasus
Ur		Urals
WS		Western Siberia
ES		Eastern Siberia
FE		Far East
	Ukraine	
DD		Donets-Dnieper
SW		South west
S		South
	Other Union Republics	
BLT		Baltic
B		Belorussia
M		Moldavia
TC		Transcaucasia
Ka		Kazakhstan
CA		Central Asia

122

Cities (in regions ordering)	Region	Population size, 000s 1986	Size change rate (%) 1979-84	1979-86
Type P$_{00}$ – restricted, both growth and development (92 cities)				
Arkhangelsk	NW	412	5	7
Tula	C	534	3	4
Ivanovo	C	476	3	2
Kalinin	C	442	6	7
Kostroma	C	273	5	7
Andropov	C	252	4	5
Yaroslavl	C	630	4	6
Podolsk	C	208	3	3
Lyubertsy	C	162	4	1
Kolomna	C	158	5	7
Novomoskovsk	C	147	0	0
Kovrov	C	155	6	8
Mytishchi	C	151	6	7
Serpukhov	C	142	1	1
Elektrostal	C	149	6	7
Kaliningrad	C	144	6	8
Orekhovo-Zuyevo	C	136	3	3
Noginsk	C	121	2	2
Gor'kiy	VV	1,409	4	5
Kirov	VV	415	4	6
Dzerzhinsk	VV	277	6	8
Yoshkar-Ola	VV	236	13	17
Tambov	CC	306	7	13
Yelets	CC	117	3	4
Volgograd	PV	981	4	6
Saratov	PV	907	4	6
Astrakhan'	PV	503	6	9
Syzran'	PV	173	4	4
Novokuybyshevsk	PV	111	1	2
Grozny	NC	399	4	6
Ordzhonikidze	NC	308	8	10
Taganrog	NC	291	5	5
Shakhty	NC	223	5	7
Novocherkassk	NC	187	3	2
Armavir	NC	170	4	5
Novorossiysk	NC	177	9	11
Maykop	NC	142	8	11
Novoshakhtinsk	NC	106	2	2
Chelyabinsk	Ur	1,107	5	7
Magnitogorsk	Ur	425	4	5
Nizhniy Tagil	Ur	423	4	6
Orsk	Ur	270	7	9
Sterlitamak	Ur	245	8	11
Zlatoust	Ur	205	3	4
Kamensk-Ural'skiy	Ur	202	6	8
Miass (Type P$_{00}$)	Ur	162	5	8

123

Appendix 3.1 Continued

Salavat	Ur	151	7	10
Kopeysk	Ur	99	−1	−32
Pervoural'sk	Ur	138	5	7
Serov	Ur	103	1	2
Novokuznetsk	WS	583	6	8
Barnaul	WS	584	7	11
Prokop'yevsk	WS	276	3	4
Biysk	WS	228	6	8
Rubtsovsk	WS	167	4	6
Leninsk-Kuznetskiy	WS	167	4	27
Kiselevsk	WS	127	3	4
Belovo	WS	117	4	4
Anzhero-Sudzhensk	WS	111	5	6
Krasnoyarsk	ES	885	8	11
Angarsk	ES	259	6	8
Khabarovsk	FE	584	8	11
Ussuriysk	FE	157	6	7
Donetsk	DD	1,081	4	6
Zaporozh'ye	DD	863	8	11
Krivoy Rog	DD	691	5	6
Zhdanov	DD	525	3	4
Voroshilovgrad	DD	503	6	9
Makeevka	DD	453	3	4
Gorlovka	DD	343	1	2
Dneprodzerzhinsk	DD	275	7	10
Kramatorsk	DD	195	6	10
Melitopol'	DD	172	5	7
Nikopol'	DD	156	6	7
Slavyansk	DD	143	1	2
Kommunarsk	DD	125	2	4
Lisichansk	DD	123	1	3
Konstantinovka	DD	114	2	2
Stakhanov	DD	110	2	2
Krasnyy Luch	DD	111	4	5
Chernovtsy	SW	249	9	14
Simferopol'	S	333	9	10
Daugavpils	Blt	126	6	9
Orsha	B	120	5	7
Kutaisi	TC	217	8	12
Sukhumi	TC	128	9	12
Karaganda	Ka	624	7	9
Semipalatinsk	Ka	324	8	14
Ust'-Kamenogorsk	Ka	313	10	14
Petropavlovsk	Ka	229	7	11
Gur'yev	Ka	147	8	12
Kokand	CA	169	6	10

Appendix 3.1 Continued

Type P_{01} – restricted growth and significant development (28 cities)

Leningrad	NW	4,359	6	7
Moscow	C	8,527	6	9
Voronezh	CC	860	7	10
Kuybyshev	PV	1,267	3	4
Kazan'	PV	1,057	5	7
Rostov	NC	992	5	6
Krasnodar	NC	615	8	10
Sverdlovsk	Ur	1,315	6	9
Per'm	Ur	1,065	5	7
Novosibirsk	WS	1,405	6	7
Irkutsk	ES	606	7	9
Vladivostok	FE	608	7	11
Khar'kov	DD	1,567	6	9
Dnepropetrovsk	DD	1,166	7	9
Kiev	SW	2,495	12	16
L'vov	SW	753	9	13
Odessa	S	1,132	6	8
Riga	Blt	890	5	7
Vilnius	Blt	555	11	15
Tallin	Blt	472	7	8
Kaunas	Blt	410	8	11
Tbilisi	TC	1,174	7	10
Baku	TC	1,114	6	9
Yerevan	TC	1,148	9	13
Alma-Ata	Ka	1,088	15	20
Tashkent	CA	2,077	12	17
Frunze	CA	617	11	24
Ashkhabad	CA	366	11	17

P_{10} – significant growth and restricted development (59 cities)

Murmansk	NW	426	8	12
Cherepovets	NW	309	11	16
Vologda	NW	273	11	15
Petrozavodsk	NW	259	7	11
Severodvinsk	NW	234	14	19
Pskov	NW	197	7	12
Bryansk	C	437	8	11
Orel	C	331	7	9
Smolensk	C	334	9	12
Vladimir	C	336	10	14
Kaluga	C	302	10	14
Lipetsk	CC	456	11	15
Penza	PV	532	8	10
Ul'ynovsk	PV	566	13	22
Engels	PV	180	9	24
Balakovo	PV	184	16	21

125

Appendix 3.1 Continued

Sochi	NC	313	7	9
Izhevsk (Ustinov)	Ur	620	10	13
Kurgan	Ur	348	9	12
Berezniki	Ur	198	4	7
Omsk	WS	1,122	8	11
Kemerovo	WS	514	7	9
Tyumen'	WS	440	15	23
Chita	ES	342	9	13
Ulan-Ude	ES	342	10	14
Bratsk	ES	245	10	14
Norilsk	ES	181	2	0.5
Komsomol'sk	FE	309	10	17
Blagoveshchensk	FE	199	12	16
Yuzhno-Sakhalinsk	FE	163	11	16
Nakhodka	FE	152	11	14
Poltava	DD	305	6	9
Kremenchug	DD	227	6	8
Berdyansk	DD	131	6	7
Kirovograd	DD	266	9	12
Kherson	S	352	7	10
Kerch	S	170	6	8
Klaipeda	Blt	197	9	12
Vitebsk	B	340	11	14
Bobruysk	B	227	14	18
Baranovichi	B	152	12	16
Tiraspol'	M	166	14	19
Bel'tsy	M	151	14	21
Kirovabad	TC	265	11	14
Leninakan	TC	226	6	9
Kirovakan	TC	167	11	14
Batumi	TC	133	6	8
Chimkent	Ka	379	12	18
Dzhambul	Ka	308	13	17
Tselinograd	Ka	269	9	14
Temirtau	Ka	226	5	6
Aktyubinsk	Ka	239	17	25
Uralsk	Ka	197	13	18
Kustanay	Ka	207	16	25
Kzyl-Orda	Ka	185	15	19
Andizhan	CA	281	16	22
Namangan	CA	283	17	25
Chirchik	CA	156	11	18
Leninabad	CA	153	13	18

P_{11} – significant growth and development (42 cities)

Novgorod	NW	224	16	20
Syktyvkar	NW	218	26	27
Ryazan'	C	500	8	10

Appendix 3.1 Continued

Kaliningrad	BLT	389	6	10
Cheboksary	VV	402	1	31
Saransk	VV	315	14	20
Kursk	CC	426	10	14
Belgorod	CC	286	14	19
Togliatti	PV	610	15	22
Volzhsky	PV	250	14	20
Stavropol	NC	299	11	16
Makhachkala	NC	311	17	24
Nal'chik	NC	231	7	12
Ufa	Ur	1,077	8	11
Orenburg	Ur	527	12	15
Tomsk	WS	483	11	15
Petropavlovsk	FE	248	12	15
Yakutsk	FE	184	15	21
Sumy	DD	262	10	15
Vinnitsa	SW	375	12	19
Zhitomir	SW	282	11	16
Chernigov	SW	285	13	20
Cherkassy	SW	280	17	23
Rovno	SW	226	20	26
Khmel'nitskiy	SW	223	22	30
Belaya Tserkov'	SW	187	17	34
Ivano-Frankovsk	SW	218	33	45
Nikolayev	S	493	9	12
Sevastopol	S	345	11	15
Minsk	B	1,510	14	20
Gomel	B	478	18	25
Mogilev	B	351	15	21
Grodno	B	255	23	31
Brest	B	230	21	30
Kishinev	M	643	20	28
Sumgait	TC	228	15	20
Pavlodar	Ka	322	13	18
Dushanbe	CA	567	9	15
Samarkand	CA	380	8	−20
Bukhara	CA	214	10	16
Fergana	CA	199	8	13
Osh	CA	204	15	21

Chapter four

What kind of solutions are in stock?

The new reforms in planning which Soviet leader Gorbachev is trying to introduce on Russian soil may look like innovations, but they are actually not. Practically all of them are borrowed from the experiences in Eastern European countries. During 1987 and 1988 the Soviets were discussing in their newspapers the possibility of giving land to the peasants, since collective farms have proved themselves ineffective. The East Europeans did not discuss it, they did it. Mr Gorbachev is going to the people, asking what would they think if they had their own land. The people from the country who are buying their daily bread in the cities know exactly what they are thinking, but they are not so sure what they are expected to think and to say.

In the Soviet Union people only now have access to such a dream as private enterprise. Long ago in Hungary life for the people began after 5 p.m., when they started moonlighting.

The Soviet realities discussed in the previous chapters do not look successful. What kind of solutions are in stock? Let us look at how the Eastern European countries handle their urban planning. Maybe their experience in this field is worth borrowing for the Soviets.

In matters of space economy a very solid expertise is available outside the Soviet borders. Much of the foreign experience in this field may be easily learnt by Soviet experts by consulting, for example, experiences accumulated in Poland. Its schools of Geography are mature; they communicate eagerly with the rest of the world; their thinking about location constraints is well-developed because of Poland's misfortunes of being located between Russia and Germany.

Theories about better organization of national urban systems are in high demand among Soviet planners, and the years of my research career in the Soviet Academy of Sciences gave me many occasions to observe how frequently the clearing house in this field was accepted to be in Poland (Borchert 1980; CNSS, 1980; Vriser 1980). An example of it was when the main journal of Soviet geographers, *Izvestiya AN SSSR, (Seriya Geografiya)* gave its front pages to a Progress Report about macro-theories applicable to urban network management (O. Medvedkov, 1980). It was very much because of the Polish origin of the macro-theories, and I find it highly curious that practically nothing is outdated and replaceable in that stock of ideas after a decade of further Soviet writings on the same subject.

The benefit of the macro-theories is not just in replacing a bulky collection of writings. In keeping with the style of main sections of this book there is statistical support for the generalities. These statistics permit us to discuss the theories while casting light, simultaneously, on controversies about cities, and the factors that change them.

Three macro-theories

Our interest in basic geographic theory leads to selection of particular theories that would fit nations with command-type economies. We find three basic approaches, best stated by the Polish scholars (Dziewonski, 1975; Leszczycki, 1975; Malisz, 1975). Each of them advocates a distinctive type of change in the spatial mechanism of intra-systemic interactions among cities. Each approach is so general that it is proper to speak about three macro-theories: (L), (D) and (M).

The first macro-theory (L) describes a monocentric organization of territory. In terms of settlement patterns, the chief role in this type of organization is played by radial linkages that converge in the most important population centre. Lesser places are subordinated to the principal centre. As a general rule, this pattern develops from benefits of scale in a main centre, with noticing just those benefits that fade slowly with distance from the pole of growth.

The second macro-theory (D) describes a polycentric organization of territory. It examines spatial structures shaped by several urban agglomerations. The benefits of agglomeration stay in limelight in this case; they spread in a space with a limited radius, and they have rapid decay or a steep gradient.

The third macro-theory (M) is concerned with a zonal-corridor type of spatial organization, in which the focus is on mainline linkages between the most active cities. It employs the possibility of keeping most of the linkages in bundles, according to the capacity of existing mainlines. The mainline linkages may be of the express type: although there are few stops, services for the roads and the circulating stock of vehicles create a belt of development anyhow.

If J represents the impact of an active mass, say, the output of electrical power in a city, and R represents distance from that active mass, then all three macro-theories postulate that:

$$\frac{dJ}{dR} < 0 \qquad (4.1)$$

The flow does not increase with distance in this case.

The model for Dziewonski's conceptual approach (D) includes:

$$\frac{dJ}{dR} < 0; \quad \frac{d^2J}{dR} > 0 \qquad (4.2)$$

i.e., the absence of an asymptote for the curve of the graph as it approaches the distance axis.

The model according to Leszczycki's ideas (L) requires, in contrast, that:

$$\frac{dJ}{dR} < 0; \quad \frac{d^2J}{dR} < 0 \qquad (4.3)$$

It is damping of flows with distance from the source, up to the asymptotic situation.

Finally, the Malisz approach (M) introduces the anisotropy of space, in which:

$$J = f(R, a)$$

where a is the azimuth for the direction of bundled traffic flows. A corridor with flows has

$$\frac{dJ}{dR} < 0; \quad \frac{d^2J}{dR} < 0$$

while outside the corridor one may observe

$$\frac{dJ}{dR} < 0; \quad \frac{d^2J}{dR} > 0$$

The use of the symbols shows that the three macro-theories complement each other completely. The range of alternatives for flows that do not grow with distance from active masses is covered in its entirety. It is therefore extremely unlikely that any centrally planned change of settlement system could be outside the basic propositions of the macro-theories.

Supporting statistical analysis

Tracing real trends attributable to (L), (D) and (M) is a matter of empirical studies. It is desirable to know, of course, the 'weight' expressing the applicability of any of the three macro-theories. In principle, none of them risks being rejected. The 'weights' would simply measure one of the three components of the structuring of geographic space, and these components are of such a general nature that they would inevitably be reflected in settlement patterns.

For empirical tests let us accept the following notation:

L – component in interaction of cities that assigns the predominant role to the national capital and gradual damping with distance.

D – component of local linkages around large cities (within agglomerations); in particular, it reflects the existence of daily commuting within a radius constrained by the daily rhythm of human activities.

M – component for flows in communications corridors that exist in accordance with the national specialization of production nodes.

Dependable results are more likely if we use indicators for L, D, M in comparisons of several urban networks. For reasons explained in Chapter 1 Ukrainian data are selected for comparisons with Eastern European national urban networks.

Our tests deal with twenty-five cities in the Ukraine, seventeen in Poland and fifteen in East Germany. These correspond to the capitals of *oblasts* in the Ukraine, pre-1975 *voivodships* in Poland and *bezirk* in East Germany. In all cases, the role of the administrative centre coincides with the main economic node in the particular *oblast, veovodship* or *bezirk*.

To judge the significance of the components L, D, M, let us uncover a response to them in T – the growth rate of cities. We will consider the intervals of time sufficiently long to eliminate short-term fluctuations, and periods in history of the territories when command-type planning operated in a set way, yet without present-day disillusionment. In the case of the Ukraine, T covers the intercensal period 1959 to 1970; for Poland it is 1960 to 1971, and for East Germany 1964 to 1973. Each T value, taken by itself, is an attribute of a particular city, but not of the urban system as a whole. The T values therefore had to be combined in some way to identify tendencies for the entire urban systems. This combining was achieved by using the regression model.

$$T = f(L, D, M). \tag{4.4}$$

The adequacy of formulae like this one may be easily tested by the techniques of the analysis of variance (AOV), well known in statistical analysis. The tests, basically, take into account how large and numerous are deviations of particular cities from a trend expressed by a regression line. The sum of squares serves to merge all deviations. This sum for the deviations from the regression line is further compared with the sum of squares associated with the slope of the regression line. There are adjustments for the cumbersomeness of the regression formula and for the total number of observations used in the regression. This yields the well know F criterion (Fisher test).

The F test was used in such a way that the adequacy of any given formula was determined in the form of a test of a statistical hypothesis was a significance level of 0.05.

The L indicators reflect the likelihood of long-distance linkages for each of the cities in the system:

R – rail distance to the national capital
V – the population potential calculated by the formula

$$V = \Sigma \left(\frac{H_i}{R_{ij}} \right) \tag{4.5}$$

where H_i is a population of the ith city; R_{ij} is the distance from i to j.

It might be recalled, that V is generally used to measure the advantages of a geographic location for urban settlement. Here we need to learn the volume of clientèle attached from outside the national centre and involved in its orbit of operations. Assuming that R_{ii} equals infinity in the formula for V we calculate V in the form of the induced potential (Lipets and Chizhov, 1972). A desirable level of informativeness of V emerges when calculations encompass all cities exceeding 20,000, e.g., 175 in the Ukraine, 70 in Poland (all *powiat* seats) and 105 in East Germany.

The D indicator reflects the likelihood of short-distance linkages for the administrative centres of *oblasts, voivodships* and *bezirk*. It is natural to make use of the population potential again, this time considering all cities of more than 20,000, but in only one particular administrative area. We shall call this variable V_1. The formula for calculating V_1 is the same as used for V, except that R_{ii} equals 1. This formula ensures calculation of the full potential based on the attractive power of the administrative capitals.

The M indicator, suggested by graph theory applications in geography, is the Konig number, known also as a measure of deviation or eccentricity (Haggett and Chorley, 1969). We use notation W1 for the metricized Konig number, derived by the formula:

$$W1 = (K_i * R_{ij})/(H_i * H_j) \tag{4.6}$$

where K is the maximum number of cities in a system that must be passed on the shortest route from a given city to the system's most distant city; R_{ij} is a nearest-neighbour distance along roads of the highest or next highest category (in km); H_i is the population of the city for which W1 is calculated, and H_j is the population of its nearest neighbour (both in thousands). Attempts to use non-metricized graph-theory indices were unsuccessful.

We had to calculate 150 regression formulae to determine the contribution made by the various components. To progress towards better formulae, we relied on the method of backward elimination of predictors. It is similar to the algorithm of step-wise regressions (Dreyper and Smith, 1973).

The following steps were undertaken:

1. The most cumbersome regression equation was computed, showing the linear relationship between T and the four indicators; the coefficient of determination was also computed.

2. The partial contribution of each indicator to the overall response expressed with the aid of the regression formula for T was measured through F-test values; consulting statistical tables for F-test, we selected a reasonable level of risk, below which the regression formula should be rejected.

3. We eliminated one of the four indicators and performed all the computations once again for the remaining three.

4. Steps 1–3 were repeated until the computer provided the answer for the remaining, most significant, indicator capable of giving an adequate explanation of tendencies in changes of T within the urban study system.

5. We compared the decline in the coefficient of determination

with changes in the *F*-test to check the elimination of the least effective indicator.

6. The indicators were rearranged to put another one in first place.
7. Steps 1–6 were repeated until the weighing of all three components for the given urban system was completed.

It may be useful to take the reader step by step through the entire sequence of regression formulae and the indications, expressed in the *F* distribution, of the adequacy of any particular proposition and of the significance of its indicators. The Polish urban system, consisting of seventeen centres, is taken as an illustration.

Step 1

The combined impact of the components L, D, M on the rate of growth of cities would be:

$$T = 0.5948 + 0.21496 W1 - 2.321\,(E{-}4)\,V_1 - 1.227$$
$$(E{-}4)V - 2.037(E{-}4)R \tag{4.7}$$

This formula explains 71 per cent of the variance of *T*, as suggested by the coefficient of determination $R^2 = 0.707$. The adequacy of the formula is quite high, as shown by the criterion *F* = 7.234, which exceeds the critical $F_{tab} = 3.26$ (for a 5 per cent risk of unjustified confidence in the formulae with four predictors of *T* and with seventeen initial observations). The order in which the components L, D, M are represented in the formula by their indicators does not affect the values of $R^2 = 0.707$ and $F = 7.234$; this fact becomes evident only when a particular predictor is excluded. The reason is the correlation among predictors. Multi-collinearity complicates the calculations, but its impact on the accuracy of the regression coefficients is weakened by statistical normalization of all the input variables.

Step 2

Next we eliminate from the formula the predictors *R* and then *V*, since their residual share of accounting for the variance of *T* is small, as suggested by the values of the *F* specific to *R* and *V*. For *R* it means, in the formula with four predictors, $F_r = 1.36$, and after elimination of *R*, when *V* is in the last place, its partial *F* value is $F_v = 2.43$. We thus obtain a formula that reflects the impact of only two macro-theories (M, D) on the growth rate of cities:

$$T = 0.4125 - 0.8264 \, W1 - 3.242 \, (E-4) \, V_1 \qquad (4.8)$$

This formula accounts for 61 per cent of the variance of *T*, with R^2 = 0.6125, and the adequacy of this more compact explanation of tendencies in *T* is much greater than before: $F = 11.07$ (compared with the critical value of $F_{tab} = 3.74$).

Step 3

A still more compact modelling of *T* comes upon excluding the predictor W1. Even though this indicator ranks first in its individual ability to model *T*, it received the lowest partial *F*, as soon as W1 faces competition from variables which signal components M and D: $F_{w1} = 7.91$ compared with $F_{v1} = 14.23$. We thus obtain the formula reflecting the contribution of the D component as well as the indirect impact of the components L and M as reflected in the D component:

$$T = 0.4346 - 3.4945(E-4) \, V_1 \qquad (4.9)$$

In this case 60 per cent of the variance of *T* is accounted for, with $R^2 = 0.6028$. The adequacy of the formula is remarkable because $F = 22.76$ – more than four times the critical $F_{tab} = 4.54$. In applied regression analysis, according to Dreyper and Smith, such an excess allows one to recommend the formula for predictive calculations (Dreyper and Smith, 1973). The fact that D-structuring of settlements (as reflected by V_1) allows for predicting is of practical significance.

Results

The results of the steps described above are in Table 4.1, but more important for our conclusions is the product of 'second enrichment': Table 4.1. All tests model *T* by regression formulae specified according to three macro-theories. All formulae but one in Table 4.1 are adequate, as indicated by the *F*-criterion. Only the last row lists negative results, suggesting the low weight of the M component, as evidenced by the indicator W1, for the East German urban system.

How good are (L), (D), (M) in specifying adequate regression formulae? We will address that question here although our main interest is to uncover the complexity of urban systems. Looking at Table 4.2 we find that for the Ukraine, the high adequacy of *T* modelling can be obtained only by the combined application of all three macro-theories. For Poland, on the other hand, the

Table 4.1 Established contribution of L, D, M components to the growth of cities in the Ukraine, Poland and East Germany

Components ranked by importance	Indicators of components and complexity of regression	R^2	F	Adequacy F_{tb}
	Ukraine (system of 25 oblast capitals)			
M, D, L	$T = f(W1, V_1, V, R)$	.667	9.99	2.87
M, D, L	$T = f(W1, V_1, R)$	.641	12.5	3.1
M, D, L	$T = f(W1, R)$	.318	5.13	3.44
M	$T = f(W1)$	.258	7.95	4.28
D, L, M	$T = f(V_1, V, R, W1)$	.667	9.99	2.87
D, L, M	$T = f(V_1, R, W1)$	.641	12.5	3.1
D, L	$T = f(V_1, R)$	.318	5.14	3.44
D	$T = f(V_1)$	.294	9.54	4.20
L, M, D	$T = f(V, R, W1, V_1)$	.667	9.99	2.87
L, M	$T = f(V, R, W1)$	.514	7.39	3.03
L	$T = f(V, R)$	.318	5.14	3.44
L	$T = f(V)$	.294	9.54	4.28
	Poland (system of 17 voivodship capitals)			
M, L, D	$T = f(W1, V_1, V, R)$	.707	7.23	3.26
M, D	$T = f(W1, V_1)$	.612	11.07	3.74
D	$T = f(V_1)$	.603	22.76	4.54
L, D, M	$T = f(V, R, V_1, W1)$	.707	7.23	3.26
L, D	$T = f(V, R, V_1)$	.707	10.42	3.41
L, D	$T = f(V, V_1)$	.605	10.41	3.74
L	$T = f(V)$	.596	22.11	4.54
M	$T = f(W1)$	.265	4.11	3.68
	East Germany (system of 15 bezirk capitals)			
L, M, D	$T = f(V, R, W1, V_1)$	.604	3.82	3.48
L, D	$T = f(V, R, V_1)$	.603	5.58	3.59
L	$T = f(V, R)$	.600	9.02	3.89
L	$T = f(V)$	.582	18.17	4.64
M, D, L	$T = f(W1, V_1, V, R)$	6.04	3.82	3.48
M, D	$T = f(W1, V_1)$	.454	4.99	3.07
D	$T = f(V_1)$	.440	10.31	4.28
M	$T = f(W1)$	.179	2.85	3.84

Note: In assessing the predictive capacity of the formulae shown in Table 4.1, we used the following proposition, which probably belongs to J. M. Wetz, who worked under Box. For an equation to be regarded as a satisfactory predictor (in the sense that the range of response values predicted by the equation is substantial compared with the standard error of the response), the observed F ratio should exceed not merely the selected percentage point of the F distribution, but be about four times the selected percentage point (Dreyper and Smith, 1973).

combined use of all three theories would make the derivation of compact formulae difficult. In this case, formulae characterizing the contribution of L or D separately assure forecasts of equally high adequacy. For East Germany, too, there is better modelling of urban growth when the predictors are specified according to the advice of either of the macro-theories; however, (L)-theory used alone also may suggest predictors which guarantee sufficient modelling.

The East German system of cities seems to be strongly patterned by dominance of its two main centres (East Berlin and Leipzig). It has an indistinct corridor structure. In Poland there is presence of patterning of urban growth both by dominance of the capital and by well-ranked centres of agglomerations. The two lines of patterning are interlinked, which happens quite logically because the ranking of centres of agglomeration occurs under influence of Warsaw. However, the most interesting characteristics of these urban systems are ahead, revealed by some additional tabulation of results. For the Ukrainian urban network, we see that the growth pattern is little affected by interlinking of (L), (D) and (M) components in the spatial structure of the urban network. This finding might easily tempt more experiments by planners to introduce partial improvements, anticipating that side-effects are limited.

From Table 4.1 we extracted information about the overlapping impact of the components L, D, M on city growth. These newly arranged results appear in Table 4.2. The bottom section may be interpreted as a measure of the complexity of urban systems. It may be seen as the level of side-effects, surprises and unavoidable failures in the partial improvements of urban settlement.

Table 4.2 Interlinking of components (L), (D), (M) in their impact on urban growth

	Ukraine	*Poland*	*East Germany (%)*
Percentage of T explained by indicators for L:	32	60	60
Percentage of T explained by indicators for D:	29	60	44
Percentage of T explained by indicators for M:	26	27	18
Percentage of T explained by all three combined:	67	71	60
Amount of overlap in the sum of explanations:	20	76	62

137

It should be made clear that the percentages of Table 4.2 characterize the coverage of the variance of *T* in the particular urban system. The most perfect modelling would lead to 100 per cent and by combining explanations, as suggested be (L), (D), (M) it cannot be more than 100 per cent for Poland and East Germany. It's always greater than elements in the fourth row, which is the actual maximum success of *T* modelling with the advice of all three macro-theories. Clearly, the sum of the three separate explanations suggested by the indicators for L, D, M would be 87 per cent for the Ukraine if there were no overlap. But we know that the combined coverage of the variance by the indicators for L, D, M is not 87 per cent but 67 per cent (32 + 29 + 26 minus overlap). Hence our ability to calculate the values that make up the lowest row in the table.

The information regarding overlap once again points out the differences in the organization of the three urban systems. For the Ukraine, the overlap is minimal; for Poland, it is quite high, and for East Germany, only slightly lower. This result has obvious practical relevance.

As a matter of fact, whenever there is a large overlap in the impact of the components on city growth, any action that modifies one of the components is likely to be weakened by the substantial remaining impact of the two other components. There would obviously be great differences in the internal inertia of the three urban systems if isolated efforts were made to modify the components. The inertia would be high in Poland, slightly lower in East Germany, and relatively low in the Ukraine. In the Ukrainian cities, it would limit risky attempts at modification to just one of the three components, but this would certainly not be justified in Poland and East Germany, where co-ordinated modifications in all three components would be required.

Discussion of the results

We now come to the crucial interpretation of the results as they apply to the control of settlement patterns. To what extent can city growth be controlled by enhancing or weakening the components L, D, M?

We obtained in Table 4.2 a concrete numerical expression for the complexity of urban systems: for the contrast in impact on the part of the components, and the overlap in impact. We also obtained proof that the two different approaches were needed to unveil the control of urban systems. The first approach, which would be appropriate for the Ukraine, but not for Poland or East

Germany, involves isolated modification of one of the three components. Such a strategy is simpler and allows greater manoeuvrability of resources, which can be focused on one particular sector.

As an illustration, we can point to the relative simplicity of measures that would enhance the D component in an urban system. Effective measures might include improvements in urban transport, manoeuvring with construction sites, and encouragement of interplant linkages within urban agglomerations. In the case of the Ukrainian urban system, such simple measures are likely to yield the same results as far more costly and drawn-out measures (say, the provision of a denser road network or the construction of a chain of new towns).

Visibility of levels of risk from centrally undertaken interventions into urban networks it is the main technical pay-off for treating the macro-theories with attention and statistical-analytical support. However, we placed all that into the conclusions for illustrating a more important point.

Isn't it amazing how inhuman the macro-theories are? They are the essence of the ideas in stocks. They indicate directions of thinking about steps that experts suggest for centralized planning. But they lie distant from the desires, perceptions and complaints of people who live in cities. Factors of location, upgrading of corridor structures, that or other attention to regional capitals all derives from the vocabulary of industrialists of the nineteenth century. Human capital is important for Soviet plans to enter the Information Age, but we have just witnessed how far away human capital is from the equipment of planning.

Conclusion

The reader may wish to see in plain English how my results contribute to better understanding of the Soviet realities. Another legitimate demand is to learn how these results may affect prospects for change in that nation.

The results, as I see them, are an additional dimension of the current crisis of Soviet society. This dimension grows from the structural foundations of Soviet urban life. Previous chapters show that Soviet urbanization has adopted a very peculiar course. Many unfortunate features of a very lasting character have developed. They include the badly managed spatial structure of the urban network and the regional urban life that has a number of very odd characteristics.

Some of these features are weaknesses and pathologies of urban life itself, but others are complications in the current steps in progress generated by the blueprints of the Soviet planners. Soviet urbanization marches on with many latent liabilities built into its spatial structure, and its march creates additional handicaps for the reforms started there in 1987 ('glasnost' and 'perestroyka').

This pessimistic view of Soviet urbanization may be surprising because, according to common sense, the previous record speed of the Soviet urban progress is the expression of its success. A success, however, may be on a shaky foundation. What I am claiming with my results corresponds to the accumulation of stubborn and odd structural features in Soviet urban life that turn into heavy burdens for the society. A number of facts support my diagnosis.

In the 'glasnost' period much (but not all) has come to light about the frequent mistakes in Soviet planning practice. It is well known that the Soviets implement their programme by patently strong means. The centralized planning of the command type economy and the monopoly of the authoritarian state are instruments that dictate all investments into urban industries. These instruments are more powerful than the bureaucrats who use and abuse them. In a closed society it is easy to hide mistakes, and, generally, little personal risk is taken by initiators in high positions. The bureaucrats survive better than the ambitious goals announced in the Soviet Five Year Plans. How big a mess the bureaucrats may create is no secret from Soviet citizens or outside observers.

Soviet urban structures are inevitably affected by the gradual accumulation of mistakes, arising from a constant bias. Remember that priorities in Soviet investments rarely take into account the acute needs of the people who must work in prescribed industries. Everywhere consumer goods, services, housing, and good medical care are undersupplied. The balance of industries in each city and within the urban network in general is not corrected by a market mechanism or by open public debates.

A recent speaker in Kremlin voiced, for example, doubts on the wisdom of keeping for decades the investment policy without any open debates. The following was addressed to the delegates of the nineteenth Conference of the Soviet Communist party:

> The main and the most massive social injustice in our nation
> affects the farmer. . . . For a long time we existed by robbing
> the farmer, by using the farmer's unpaid toil. And later, when
> the farmer became extinct in many regions of our nation, we
> discovered oil opportunities. To sell abroad our national
> reserves for the sake of importing food is a criminal activity of
> the creators of our stagnant period. (Aidak, 1988:5)

The odd features in Soviet urbanization, described in my results, may originate from the indicated causes. It may be applicable, for example, to the fact about a majority of bigger Soviet cities having functional profiles of 'company towns'. Or it may explain how main Soviet centres of manufacturing started to overload the national network of railroads.

My task in the book was not to discover the causes but rather to measure objectively the weight these odd features load on to Soviet urbanization, slowing its manoeuvrability.

By 'odd' I mean that a feature is inconsistent with a gradual progress of the society and its economy to a more developed phase, with better internal balances, with rational links among settlements and industries. In essence, my understanding of odd structural features is similar to the critical quotation shown above.

The characteristics I have found 'odd' are not on the surface and, for this reason, not in the current debate nor in the corrective measures. In other words, I tell about rocks and sand bars right in the course of the Soviet ship and unknown to its seamen.

The task of providing visibility for these problems has required, as this book testifies, complex analytical techniques. Bringing an unflattering image to the surface is a job that many Soviet scholars dislike and avoid. The Soviet statistical sources are of the poorest help in this field. If they communicate any message, one must be a detective to find it or construct it inferentially. It took years, and emigration from the Soviet Union, to get these results and to publish them.

Our first findings concerned the hierarchies in the Soviet urban network (Chapter 1). Ordering of cities by size provided the simplest understanding of these hierarchies, following the 'Rank-Size Rule' well known to geographers. On the level of this understanding it transpires that the Soviets are now objectively at a disadvantage because of the weaknesses revealed in Moscow's leadership among Soviet cities. Its size, as well as its urban maturity, does not place this centre high enough in the proportions suggested by the sizes of other Soviet cities.

Stated differently, Moscow does not possess unquestionable superiority over the cities it wants to rule. Politically it claims superiority, but its manpower support may be less than the claim. At least, it is less impressive to the circle of immediately sub-ordinated regional capitals who consider their own weight (millionaire cities, as a rule) in respect their own areas of dominance.

I will add shortly a number of reservations to make a more accurate interpretation of the findings. At the moment let us realize that the communicated message interferes with the conduct of the reforms announced in the Soviet Union.

The reforms come from the top, and Moscow, because of its political functions, is at the origin and at the steering wheel of the process. To put it plainly, the reforms may not have influence further than Moscow. It follows from traditions of the Soviet society, where all its official life is organized around the image of central leadership. Now we are learning that the claim for Moscow's leadership has weaknesses, whereas regional capitals have better potential for acceptance of their leadership in smaller parts of the nation. These regional challenges certainly may endanger the reforms or slow them down. Also the structure of the urban hierarchies may amplify the voices from regional capitals (as Erevan or Stepanakert show) thanks to local loyalty, so that it may be hard for the authority of Moscow to silence these voices.

All is relative and on average in the conclusions of this kind. There are exceptionally undersized regional centres too, and Kiev is an example of it. Its political leadership over the territory of the Ukraine appears to have less dominating manpower support than Moscow has, in relative terms.

I also quite willingly admit the imperfections in the yardstick applied for the task of assessing the relative dominance of cities – all based on the number of inhabitants. This measure omits regional differences in qualitative characteristics of the manpower and the percentage of people not in the labour force. The reasoning borrowed from the 'Rank-Size Rule' certainly has too many and too bold assumptions.

These limitations do not ruin the findings; rather they suggest a restricted area of applicability. When I accept that Moscow concentrates the better trained manpower it does not permit me to extend the statement about this city's weakness to real-life situations, when the leadership depends on the qualifications of the teams at work, on the brainpower deployment.

The weakness comes mostly in situations wherein only the number of people is of importance. Moscow may maintain its position of political leadership when efforts are set in order of events, and all is orderly and disciplined. The weakness comes as soon as the events turn into behaviour of the crowds, into grass-roots activism or into influences expressed by the 'silent majority' of the populace. Soviet political statements, when they serve the purpose of the populist policy, take into account the noted significance of the regional centres. Repeatedly, the press reports that either Mr Gorbachev or Mr Ligachev, his opponent, chose to defend his view not in Moscow but in Vladivostok, in Murmansk, in Gorky, etc. It is not their folly, neither it is an escape from too many watchdogs in Moscow. Instead, according to our conclusions about the Soviet urban hierarchy, each such case is an attempt to mobilize support with the amplifying authority of the place where the speech is performed. Undisputed authority of the place in the sphere of its political dominance is what attracts Soviet leaders to provincial capitals for the purpose of making political statements.

The second simply-stated interpretation arises from the findings on the interdependence between the urban hierarchy and presence of development on all national territory, also discussed in Chapter 1. I have found that the ideas about keeping development in the well-defined course, with orientation on better balances and rationality, contradict the reality studied. To put it bluntly, I have tested to see if the centralized planning does what it promises, and the outcome is negative. The failure is not of the type frequently acknowledged now in the Soviet press (Aganbegyan, 1988): it is more serious than just some isolated under-fulfilment of quotas for producing steel, coal, butter or boots.

I tested the presence of consistency for upgrading in parallel the cities and the rest of the economic space in the nation. The evidence revealed a record of wanderings, of steps made as if at random and blindly, with frequently changed directions.

This evidence appeared in the test of the Soviet union, a separate test of the Ukraine, and confirmatory examples of the Eastern European nations. I want to be very fair. It could happen, for example, that the Soviet Union is too big and too different in its many regions for implementing one definite line of improvements for its spatial structure. If that were the case, then success might come for smaller entities. The Ukraine is furnished with its own mechanism of the centralized planning, stationed in Kiev. The same is true of capitals of all studied nations of Eastern Europe. I tried to find any consistently stable line of improvements by taking data for the most untroubled decades of the period following the Second World War.

Centralized planning has its strong points: it could implement deep reorganizations in economy of the Soviet Union. The New Economic Policy with its market economy was successfully dismantled. Industrialization made its speedy progress. The war economy successfully provided the Soviet Army. The Soviets were the first to launch a space probe and to dispatch man to space. They are quite competitive in their build-up of the military – industrial complex. The programme of providing mass housing units was also a success in the Khrushchev years and long after.

I have examined in this book the planners' ability to make complex and well-co-ordinated improvements. Are they capable of implementing a development with rational interconnected functions of cities and of spaces outside the urban network? This, after all, is the very area where, some theorists predict, centralized planning has specific superiority. The only evidence from my tests show wanderings of the planners.

The announced reforms must deal with the wanderings legacy. There is little rationality in the spatial organization of the national territory by interrelated functions of the cities. The planners may know what is more rational, but giving them all benefits of the doubt, they did not have the power to implement their knowledge in the 1960s or 1970s, years free of the current difficulties of the Soviet economy. In those years the planners were not passive at all: they set forward ambitious goals of restructuring 666 'regional and local settlement systems' (Khodzhaev and Khorey, 1978). There was little success in this direction, however. The fiasco seems to be well-realized both by the Soviet and foreign experts (Fuchs, 1983; Listengurt and Portyanskiy, 1983; Demko and Fuchs, 1984; Dienes 1987). With that experience it is unwise to expect a sudden miracle from the Soviet planners. Additional reasons for the scepticism are provided by my findings. They show that the Soviet Union now faces difficult coexistence, this time not with the USA but with its own structuring of the economic space.

If the foregoing interpretation of the findings in Chapter 1 sounds too general, it is because the findings are exactly of that kind. Our approach to socio-economic realities focuses on aggregations that cities are made of. Our interest is on forces of organization represented by cities without any breakdown to the functions that may be very different for particular cities. Remember that our portrait of urbanization must cover the largest national territory on the present-day political map of the world: by necessity we must start with a very schematic outline.

The task of looking inside each city comes in Chapter 2. A collection of functional profiles comes to light. In this shortest form, the findings are shown on the maps of cities dependent on distant linkages. But maps have a language of their own, and they should be explained. Stated plainly, Soviets are now overtaxed by the dated creations of their industrialization. The cities dependent on abnormally costly linkages cover as a network all the nation, one-sixth of the land in our planet.

By employing the approach of typologies I have uncovered a striking fact: the Soviet Union is a nation of company towns. Their peculiarity is in letting such centres grow to sizes well above 100,000. Their other peculiarity is the existence of such cities prohibitively distant from one another.

Practically all major centres carry the legacy of basic industrial enterprises. Manufacturing is patterned by fabricating first of all the means of production', and very parsimonious provision of consumer goods. As a result of that policy of building industrial giants, the Soviets are doomed to maintain most of their industry by expensive inter-regional trade. Internal distances are comparable to those of international trade of the Western nations. Flows are heavy because each city brings from another corner of the nation the lion's share of its semi-products of raw materials necessary for fabricating its share of heavy industrial equipment. The interchange goes by land transportation, with very little help from shipping.

The real cost of transportation in the Soviet Union is bigger than Soviet statistics show. Freight rates are quite artificial and ridiculously low. The situation amounts to constant outlay of subsidies from the consumers' pockets into the accounts of the Soviet industrial giants that never went out of business, up to the end of 1988, never mind how bankrupt they are. At the end of 1988 a spectacular display of bankrupt State enterprises had occurred in the Soviet Union. This provides additional support to our findings. And the oversized producers of 'means of production' are prominent in the march of bankrupt enterprises (for example, the integrated iron and steel works of Rustavi, Georgia, was the first sizeable case).

Chapter 2 also shows that the locational pattern of industrialization already in place in 1970 continues to dominate the functioning of Soviet urban centres. Cities with a prominent orientation towards services or information processing are so rare that I could not find an impact from the Communication Age on the spatial structures of the Soviet urban life under study here. The list of factors with a negative influence on the announced reforms had certainly been lengthened by the results uncovered in Chapter 2.

Chapter 3 has equally pessimistic material. It follows the study design adopted in the whole book: step-by-step uncovering of man-made spatial structures of urbanization, those capable of influencing the present life of the society.

Chapter 3 compares the tendency to add to the Soviet cities what they already have 'growth' with the tendency to bring changes appropriate in the Communication Age (development). The first half of the 1980s clearly saw a decline in growth and little of development in the set of 221 main Soviet cities. This finding is not surprising. The lack of growth and development is admitted now for the Soviet Union in general, and the authority telling about it is the top political boss of that nation!

My diagnosis shows that most Soviet cities are badly affected by stagnation. If corrective steps are going to be taken, the Soviets face the task of implementing them practically in all regions. But such steps cannot be mechanistically applied, because of deep regional differences in types of stagnation.

The character of the specific regions comes to light when we look at the regional mix of growth and development. There are four types of urban dynamics in that mix (my designations for them are: 00, 01, 10, and 00), and I have found that each type exists in the most tight association with such economico-geographical features as clustering of urban settlements, their connectedness, presence or absence of rural vitality inside the regions, and the degree of marginality experienced by the region in the national territory.

Urban stagnation clearly varies according to the presence of the enumerated features in the Soviet economic regions. For the purpose of my explanatory comments, we may dismiss the possible distinctions between directions of cause-and-effect relationships – whereas urban stagnation is patterned by the features of each region or whereas these features follow the stagnation as an added weight from a logically necessary complement. I find that the associations uncover an additional and heavy complement to the stagnation of cities.

The Soviet's earlier economic successes invariably derived from industrial investments and urban construction. I have found that presently it will not be enough to make corrective steps only in the cities. The unfavourable dynamics of the cities are tightly linked with larger geographical structures, the big territories of the regions. Such links add complication to the reforms.

Finally, Chapter 4 finds certain macro-structures embracing all the national territory. They are organizational directions for the urban networks, and I have examined them by comparing the Soviet Union to the other communist nations of Eastern Europe.

One of the organizational directions cultivates direct connections between provincial cities and the national capital. It creates a systematic gradient in the density of the urban network: the spacing of cities increases proportional to the distance from the national capital.

The second organizational direction provides more uniformity in the urban network. It develops if a number of regional centres get the privilege of building direct connections with subordinated but still-significant cities, each inside its region. Next, the subordinated cities act as sub-regional centres by developing direct connections down to the closest smaller communities. Several strata of focal points may come to life in this process. Notice that the urban network develops in this case in the form suggested by the Central Place Theories discussed in Chapter 1.

The third possibility lies in beefing up the cities located along the most important main lines – usually between major urban centres on both ends of the mainline. In the Soviet Union this pattern appears, for example, in the corridor between Moscow and Donetsk. It is obvious that the third organizational direction, the one in accordance with the theory of 'urban corridors', permits us to leave huge areas outside any urban influence. In the vast Siberian space it looks like the only realistic alternative for the foreseeable future.

The results of Chapter 4 show that Soviet urbanization has followed all three organizational principles. Diversity of the Soviet regions being very deep (as Chapter 3 shows), this result was expected, of course. What is unusual is the very tight interconnectedness of spatial structures materialized according to steps one, two and three. But that structural interconnectedness also translates into the impossibility of undertaking step-by-step

147

improvements in the network of cities.

In a period of reforms it is dangerous to go for a long time without success. A chance for early success may come by concentrating all efforts on a very limited but key area of interventions. But such a strategy is next to impossible in the structure of the Soviet urban network. It is the least suitable area, according to findings in Chapter 4, for efforts if they are not of the frontal character. Limited corrective steps are not applicable here, because they are more likely to bring a mess than improvements.

If one attempts to increase the economic weight, for example, of the cities in the Moscow–Donetsk corridor, one would immediately upset the other directions in the subordination of cities: there would be an impact on direct connections with Moscow and on the performance of several regional centres in their zone of influence.

An alternative interpretation of my findings in Chapter 4 may be the following: it is highly unlikely to expect that improvements in the structure of the Soviet urban system will come in the period of the reforms. The tight structural interdependence in the network suggests that only a frontal intervention would be meaningful and bring a predicted outcome, and such an intervention requires a lot of time and resources. Right now both time and resources are at a premium for the Kremlin. Consequently, the subject of improvements for the urban network is postponed.

The view on the necessity of frontal improvements for the Soviet urban network is not only voiced in Chapter 4. The majority of the Soviet experts in urban planning expressed very much the same opinion in the 1970s, in the much-advertised plan entitled 'Nationally Unified System of Settlement'. As soon as the Soviet economy entered into a phase of difficulties, in the late 1970s, that plan was shelved to rest in peace. My findings in Chapter 4 show that the Plan cannot easily be brought back to life.

While the Soviets may not afford these improvements, they continue to be constrained by the unfavourable characteristics of their urbanization. Once the Soviet Union was a field of victories for planners, but now it suffers from weeds. Cities are the ground for making steps ahead with reforms, but in their present shape they are also a source of constant and strong impediments for all efforts of the Soviets.

Personal epilogue

As I started to write this book, I could not for a long time get rid of a double vision. I probe my subject with instruments of science and at the same time all my senses perceive the reality on which the study is based. From my previous life experience in the Soviet Union I know it deeper than instruments register.

This book attempts to uncover, with objective approaches of science, how far Soviet urbanization supports the claim of the Soviets that they steer the nation in a rational way, with centralized planning. Most of my analysis was done in Moscow. Conditions of my work in the Soviet Academy of Sciences did not allow me to state openly the reassessing character of my project, and I was not supposed to deviate from the aspects of urbanization which the Soviet planners bring to the attention of science. Although my research was constrained at least I could start it.

I worked in the research department of the Academy only four or five blocks from the Kremlin. I could see the red stars of the Kremlin every day, and I understood how dictatorial they are for the nation. The 'Thaw' of the late 1950s was a distant memory in Moscow. There was no more talk about the horrible crimes of Joseph Stalin; the dictator's portrait started returning to State offices. No sign of today's attempts to bring reforms to the Soviet society was visible.

In my enquiry there was a chance to apply techniques which Geography, a discipline with active international ties, gave me. I studied at Moscow State University in years when the library added one innovative volume after another: Quantitative Revolution, Spatial Analysis, Social Awareness, Geography of Welfare.

Later on I was lucky to have access to uncensured professional discussions. Yuri Medvedkov, my husband, and I organized seminars in our home, where we met visitors from many lands, the very authors of the milestone volumes, and discussed their findings and our results.

Personal epilogue

In 1980 my first version of the book was ready, and soon the Academy Publishers were sending it to the printer's press 'Nauka'. Then all stopped abruptly because of our dissident seminars and our insistence on free international contacts.

Since that time I engaged in six years of battles for elementary human rights. The portrait of Joseph Stalin on the wall confronted me when I was arrested by the KGB. This new experience disclosed for me many dimensions of Soviet realities outside my initial study track.

Yet I keep this book, quite deliberately, within initial limits. It is based on official statistical data, not on my experiences. I may use the experiences in another project (O. Medvedkov, 1988a). This study has the benefit of being completed in the USA, where I have the possibility for quiet thinking and means for updating information. With all the complexity of my double vision, I am pretty far from the Soviet reign of double thinking. My findings may be reported as they are, in an uncensured form. I may write openly, not between lines. One must have years of painful practice of publishing in the Soviet Union to feel as strongly as I do, what a treasure is in the First Amendment.

How does the urban hierarchy function in a nation where ranks of people are a master of great importance? How do old industries, which cities retain from early Five-Year Plans, influence their growth? Why did the Soviets come to an impasse with a run-down rail network? How does the Soviet Union, a superpower, allows a living standard for its people that nations of the Third World do not tolerate? These and similar questions must have definite answers, without interference or censure.

I know that in my book all is carefully verified, quietly measured, distilled from biases and emotions. These are the standards of my profession. I accept them. I know their value. But I have learnt too that this approach does not disclose all the urgency in the Soviet Union.

I had a hard time rethinking my study again, and in English. A word processor with a built-in thesaurus and various editing programs helped keep my attention on the substance of my results rather than on the technicalities of a language which is not native for me. This rethinking, however, helped me to have a critical vision of the results.

A personal epilogue has privileges. It is the place for my emotions, for dissatisfaction with cool instruments of research. They bring only part of the understanding: there are tragedies and sufferings of people behind the symbols of my models, and the models do not show it. My double vision is returning to me.

150

Images come to my mind with each entry of data into tables, when a familiar place name comes with it. I remember sad and tired faces, shabby clothes, stooped bodies. I see people wasting time in lines, rushing from one store with empty shelves to another, forcing their way on to crowded buses.

Behind almost each name of a Soviet city I see how hopeless is the life of its inhabitants. I remember the structure of Soviet urban centres: two or three stores to be visited daily, neglected parks and libraries, schools with kids in uniform, hospitals which look like prisons. Next, at exiting roads, real prisons stand among barracks and factories. Little differentiates the demeanour of those three institutions, inside or outside.

More colours and animation appear on the busy streets of Moscow, Leningrad, Kiev, Tbilisi, and Riga – window-dressing for foreigners. At the same time, constant haste and life on the run reign there. There is none of the drowsiness of smaller towns, but gone also is a human attitude of people to each other, Pedestrians are more aggressive than friendly.

Rushing crowds in corridors of the Moscow subway are like an avalanche. The crowd at peak hours engulfs you and drags ahead, bumping into walls and steel barriers. The same may await you in a big department store. Stampeding people are like a flow of mud, which spoils and crushes all. But they still are minor inconveniences. More sizeable tragedies are all around.

Social strata are rigid in the Soviet Union. Residents of small provincial towns are usually regarded as failures. Outside the five best cities, be prepared for extremes in shortages of food. Risk of robberies and mugging goes up, and the range of available jobs goes down. It is not surprising that young people dream of escaping into main centres. Once they do it, the burning question arises: where to find shelter?

During the 1980s there have been fewer additional housing units than additional families. Shared apartments are the fate for many newly-weds, and kitchens and corridors frequently turn into battlefields. The rate of divorce grows higher and higher.

In the vocabulary of the Soviet people, there is no equivalent for 'privacy' or 'fun'. If the first word is known to some intellectuals, the second one is completely absent: there is no fun in such a life.

True, the events of the two recent years brought another infrequently used word into the Soviet vocabulary – 'excitement'. This is not the place to judge Gorbachev's reforms and my book is not entitled to deal with it. But nevertheless it is worthwhile mentioning that life, particularly for intellectuals, became much more exiting than it used to be. It does not solve the problems of daily life and daily bread: the shortages of goods are even worse but it brought the new spirit. Nobody knows how long it may continue, but people are grabbing the opportunity of 'perestroyka' and it is becoming more difficult for the authorities to keep the 'glasnost' in their own limits. Cities are changing their appearance: in the front of the State supermarket, there are plenty of small private shops; there are private cafés and restaurants.

I will stop my emotional description of the Soviet reality on this positive note. The main text of the book casts light in another way on the topics of freedom and control in Soviet life, but I think it is helpful to have both the deep sophisticated analyses and the broad vision.

I want to express my gratitude to Jack Hollander, Vice-President of the Ohio State University, and to Charles J. Hermann, Director of the Mershon Center of the University for the research grant to complete this book. I enjoy the hospitality of the Center where I found the ecologic niche about which the scholar may only dream. I am also grateful to Wittenburg University where my lectures on Soviet urbanization and other subjects contributed much to this book.

Very substantial support has come from the National Council for Soviet and East European Research.

There was constant support and patience from my family, and it took a lot of understanding for Mike and Mary, my children, to accept that their mom is spending much more time at the PC keyboard than with them.

Much encouragement and help came from geographers in Ohio State University and elsewhere. I address my gratitude for attention, in particular, to Lawrence Brown, Galia and Guy Burgel, George Demko, Tony French, Robert Gohstand, Chauncy Harris, Julian Wolpert, and Craig ZumBrunnen. My Russian-English was much improved by the patient editing of Kesia Sproat.

Other colleagues know that I also thank them deeply, but they must be left anonymous: they have arranged for my manuscript to cross Soviet frontiers and reunite with me in Columbus. There was no way for me to take in when the Soviets, in September of 1986, gave me and all my family two days to pack and leave the country for good. To sum up, I must say that for me the development of this book symbolizes the undying spirit of solidarity in the international community of scholars.

Columbus, Ohio

References

Aganbegyan, A. (1988) *The Economic Challenges of Perestroyka*, Bloomington: Indiana University Press.

Aidak, A.P. (1988) Vystuplenie na vsesoyuznoy partiynoy konferentsii, *Pravda*, 2nd July: 5.

Alaev, E. (1986) *Social and Economic Geography*, Moscow: Progress Publishers.

Alexandersson, G. (1956) *The Industrial Structure of American Cities*, Lincoln: University of Nebraska Press.

Alperovich, G. (1984) The size distribution of cities: On Empirical Validity of the Rank-Size Rule, *Journal of Urban Economics*, 16: 232–9.

Arapov, V.M., and Shreider Y.A. (1977) Classifikatsiya i raspredelenie po rangy [Classification and Rank Distribution], *Nauchno-Teknicheskaya informatsiya [VINITI Publishing]*, 2(11–12): 10–20.

Asamy, Y. (1986) Fitting the Rank-Size Rule to legal cities, *Geographical Analysis*, 18: 243–52.

Auerbach, F. (1913) Das Gesetz der Befolkerungskonzentration, *Peterman's Geographische Mitteilungen*, 59: 74–6.

Axelrod, S.A. (1978) *Sociocultural Potential. Report in the Laboratory of Human Ecology [Unpublished]*, Moscow: Institute of Geography, Soviet Academy of Sciences.

Baily, N. (1970) *Matematika v biologii i medicine [Mathematics in Biology and Medicine]*, Moscow: Mir Publishers.

Baranskiy, N.N. (1956) *Economic Geography of the USSR*, Moscow: Foreign Language Publishing House.

Beckmann, M.J. (1958) City hierarchies and the distribution of city sizes, *Economic Development and Cultural Change*, 6: 243–8.

Berry, B.J.L. (1961) City size distributions and economic development, *Economic Development and Cultural Change*, 9: 573–87.

Berry, B.J.L. (1978) Latent structure of urban systems: research methods and findings, in L.S. Bourne, and J.W. Simmons (eds), *Systems of Cities: Readings on Structure, Growth, and Policy*, New York: Oxford University Press: pp. 220–31.

Berry, B.J.L., and Horton, F.E. (1970) *Geographic Perspectives on Urban Systems: With Integrated Reading*, Englewood Cliffs: Prentice Hall.

References

Borchert, J.G. (1980) The Dutch settlement system, in K. Dziewonski (ed.), *National settlement systems. Topical and national reports*, Warsaw: IGU Commission on National Settlement Systems, pp. 281–336.

Boventer, E. von (1973) City-size systems: theoretical issues, empirical regularities, and planning guides. *Urban Studies*, 10(2): 145–62.

Cassetti, E., King, L., and Odlund J. (1971) The formalization and theory of concepts of growth poles in a spatial context, *Environment and Planning*, 3(4): 377–82.

Chislennost, sostav naseleniva SSSR po dannym Vsesoyuznoy perepisi naseleniva 1979 goda [Soviet Census of Population: 1979] (1984), Moscow: Finansy i Statistika.

Christaller, W. (1933) *Die zentralen Orte in Suddeutschland*, Jena: G. Fisher.

Christaller, W. (1956) *Central Places in Southern Germany*, Englewood Cliffs: Prentice Hall.

Christaller, W. (1962) Die Hierarchie der Stadte, *in Proceedings of the IGU Symposium: Urban Geography, 1960* Lund: Lund University Press, pp. 3–11.

CNSS (1980) Systems of main urban centers (functioning within National settlement systems), in *National Settlement Systems. Topical and National Reports*, Warsaw: IGU Commission on National Settlement Systems, pp. 103–20.

Crowley, R.W. (1978) Labor force growth and specialization in Canadian cities, in L.S. Bourke, and J.W. Simmons (eds), *Systems of Cities. Readings on Structure, Growth, and Policy*, New York: Oxford University Press, pp. 207–19.

Dacey, M.F. (1965) The geometry of Central Place Theory, *Geographiska Annaler*, 47(2): 11–124.

Davidovich, V.G. (1976) Proiskhozhdenie i razmer migratsii naselenia [Origins and Sizes of Population Migrations], in *Problemy migratsii naseleniva v SSSR [Questions about Population Migrations in the USSR]*, Moscow: Mysl Publishers.

Dawson, A.H. (ed.) (1987) *Planning in Eastern Europe*, New York: St Martin's Press.

Demko, G.J., and Fuchs, R.J. (1984) Urban policy and settlement system change in the USSR, 1897–1979, in G.J. Demko, and R.J. Fuchs (eds), *Geographical Studies on the Soviet Union. Essays in Honor of Chauncy D. Harris*. Chicago: The University of Chicago, pp. 53–69.

Dienes, L. (1987) Regional planning and the development of Soviet Asia, *Soviet Geography: Review and Translation*, 28: 287–314.

Dzaoshvily, V.S. (1978) *Uranizatsiya Gruzii: prichini, processi, problemy [Urbanization of Georgia: Origin, Processes, Problems]*, Tbilisi: Mezniereba Publishers.

Dreyper, N., and Smith G. (1973) *Prikladnov regressionniv analiz [Applied Regression Analysis]*, Moscow: Statistica Publishers.

Dziewonski, K. (1975) The structure of population in the future settlement system in Poland, *Geographia Polonica*, 32: 3–15.

References

Dziewonski, K., and Jerczinski M. (1978) Theory, methods of analysis, and historical development of national settlement systems, *Geographia Polonica*, **39**: 201–11.

Ekkel, B.M. (1978) Polyarizatsia structury urbanizirovannykh territoriy v zelyakh uluchshenia kachestva okruzhayuschey sredy i conservatsii [Polarization of Built-Up Territories for the Purpose of Creating a Healthier Environment with Better Provisions for Conservation of Nature], *Izvestia Akademii Nauk SSSR, ser. geogr.*, **5**: 54–64.

Fano, P. (1969) City size distributions and central places, *Papers of the Regional Science Association*, **22**: 29–38.

French, R., (ed.) (1979) *The Socialist City*, New York: Wiley.

Fuchs, R.J. (1983) Mobility and settlement system integration in the USSR, *Soviet Geography: Review and Translation*, **24**: 547–57.

Giese, E. (1979) Transformation of Islamic cities in Soviet Middle Asia into socialist cities, in R. French, and J.I.F. Hamilton (eds), *The Socialist City*, New York: Wiley pp. 145–65.

Golz, G.A., and Filina, V.N. (1977) O raspredelenii grusovikh potokov na transportnoy seti [On Distribution of Freight Flows on Highway Networks], in *Trudy IKTP, Vypusk 61: Voprosy razvitiya edinoy transportnoy seti SSSR [Works of the Institute of Complex Transportation Problems. Moscow].*

Gradostroitelstvo: Compendium of the Central Urban Planning Institute (1975), Moscow: Stroitelstvo.

Grossman, G. (1979) Notes on the illegal private economy and corruption, in *Soviet Economy in a Time of Change. A Compendium of Papers Submitted to the Joint Economic Committee, Congress of the United States. Vol. 1.* Washington, DC: US Government Printing Office, pp. 834–54.

Gudjiabidze, V.V. (1974) Pravilo Zipfa v prilozhenii k sisteme rasselenia Zakavkazya [Zipf Rule in Application to Settlement Systems of Transcaucasia], in *Geografiya i Matematics [Geography and Mathematics]*, Tartu: Tartu University Press.

Hagget, P., and Chorley R.J. (1969) *Network Analysis in Geography*, London: Arnolds.

Hagget, P., Cliff, A.D., and Frey A. (1977) *Locational Analysis in Human Geography: 2d ed.* London: Arnolds.

Harary, F. (1980) *Graph Theory*, Reading, Mass.: Addison-Wesley.

Hardt, J.P.K.R.F. (1987) Gorbachev's economic plans: prospects and risks, in *Gorbachev's Economic Plans. Vol. 2 Study Papers Submitted to the Joint Economic Committee, Congress of the United States*, Washington, D.C.: US Government Printing Office, pp. vii–xvi.

Harris, C.D. (1943, January) A functional classification of cities in the United States, *Geographical Review*, **33**(1): 86–99.

Harris, C.D. (1945, January) The cities of the Soviet Union, *Geographical Review*, **35**(1): 107–21.

Harris, C.D. (1970) *Cities of the Soviet Union: Studies on their Functions, Size, Density and Growth*, Chicago: Rand McNally.

Harris, C.D. (1988) Basic geography in J. Cracraft (ed.) *The Soviet Union*

155

References

Today: An Interpretive Guide (2nd edn) Chicago: The University of Chicago, p. 141.

Hotelling, K.L. (1936) Relations between two sets of variables, *Biometrica*, **28**: 321–77.

Iodo, I.A. (1985) Planirovochnye aspekty regulirovaniya razvitiya Minska [Planning Aspects of the Regulation of Minsk's Development], *Stroitelstvo i Arkgitektura Belorussii*, **2**: 12–15.

Itogi Vsesoyuznov perepisi naseleniva 1970 goda: tom 1 [Soviet Population Census: 1970: Vol. 1] (1972) Moscow: Finansy i Statistika.

Itogi Vsesoyuznov perepisi naseleniva 1970 goda: tom 4 [Soviet Population Census: 1970: Vol. 4] (1973) Moscow: Finansy i Statistika.

Itogi Vsesoyuznov perepisi naseleniva 1970 goda: tom 2 [Soviet Population Census: 1970: Vol. 2] (1974) Moscow: Finansy i Statistika.

James, M. (1985) *Classification Algorithms*, New York: Wiley.

Jensen, R.G. (1984) The anti-metropolitan syndrome in Soviet Urban policy, in G.J. Demko, and R.J. Fuchs (eds), *Geographical Studies on the Soviet Union. Essays in Honor of Chauncy D. Harris*, Chicago: The University of Chicago, pp 71–91.

Johnston, J. (1980) *Econometric Methods (Russian edition)*, Moscow: Statistica Publishers.

Khanin, S.E. (1976) Problemy izucheniya dinamiki gorodskih poselenii [Problems in Studies in Urban Settlement Dynamics], *Urbanizatsiya i formirovanie sistem rasseleniva [Urbanization and Development of Settlement]*, Moscow: Soviet Geographical Society.

Khodzhaev, D., and Khorev B.S. (1978) The concept of a unified settlement system and the planned controle of the growth of towns in the USSR, in L.S. Bourne, and J.W. Simmons (eds), *Systems of Cities: Readings on Structure, Growth, and Policy*, New York: Oxford University Press, pp. 511–18.

Khorev, B.S. (ed.) (1980) *Problemy rassekleniya v SSSR [Problems of Settlement in the USSR]*. Moscow: Statistika.

Kilkpatrick, M. (1987) *Business Statistics Using Lotus 1-2-3*, New York: Wiley.

Lappo, G.M. (1978) *Razvitiye gorodskikh agglomeratsiy v SSSR [Development of Urban Agglomerations in the USSR]*, Moscow: Nauka.

Lavrov, C.B., Litovka, O.P., and Medvedkov Y.V. (1979) Ekologiya rasseleniya [Settlement Ecology]. In *Geografiva naseleniya v sisteme kompleksnogo ekonomicheskogo i sotsialnogo razvitiya*, Leningrad: Soviet Geographical Society, pp. 12–26.

Leszczycki, S. (1975) The growth limits of urban-industrial agglomerations in spatial development at the national level, *Geographia Polonica*, **32**: 3–15.

Levine, M.S. (1983) Possible causes of the deterioration of Soviet productivity growth in the period 1976–1980, in *Soviet Economy in the 1980s and Prospects. Part 1. Selected Papers Submitted to the Joint Economic Committee. Congress of the United States*, Washington, DC: US Government Printing Office, pp. 153–68.

Levine, M.S. (1986) *Canonical Analysis and Factor Comparisons Sage University Paper series on Quantitative Applications in the Social Sciences, 07–001*, Beverly Hills: Sage Publications.

Lewin, M. (1988) *The Gorbachev Phenomenon*, Berkley: University of California Press.

Lipets, Y.G., and N.N. Chizhov (1972) Statisticheskive metodv izuchenia potenciala naselenia [Statistical Methods for Studying the Population Potential], in Problemv Sovremennov Urbanizatssi [Problems in Contemporary Urbanization]. Moscow: Statistica Publishers.

Listengurt, F.M., and Portyanskiy I.A. (1983) Alternativy razvitiya rasselenyia v SSSR [Alternatives for the Development of Settlement in the USSR], *Izvestiya AN SSSR, er. geogr*, 4: 45–55.

Loesch, A. (1940) *Die raumliche Ordnung der Wirtschaft*, Jena: G. Fisher.

Loesch, A. (1954) *The Economics of Location*, New Haven: Yale University Press.

Lola, A.M. (1983) Existing systems of settlements in the USSR and some research problems relating to the transformation of systems, *Soviet Geography: Review and Translation*, 24: 18–30.

Malisz, B. (1975) Research work as an input in the construction of the National Plan. Appendix: The Band-Node Model of the settlement network in Poland, *Geographia Polonica*, 32: 15–27.

Matlin, I.S. (1974) *Modelirovanie raspredeleniya naseleniya [Modelling of Population Distribution]*, Moscow: Nauka Publishers.

Medvedkov, O. (1975) *Sravnitelnaya geofrafiya makrosistem gorodov [Comparative Geography of Urban Macro-systems. Dissertation]*, Moscow: Institute Geography, Soviet Academy of Sciences.

Medvedkov, O. (1976) Parametrisation of Urban Macrosystem by Techniques of Geographical Information Systems, in *International Geography-76*, Moscow: Nauka Publishers.

Medvedkov, O. (1977) Sistemnie vzaimosvyasi gorodskoy sredy kak faktor gorodskogo razvitiya [Internal Systemic Links Within Urban Environment as a Factor in Urban Change], in *Gorodskaya sreda i puti ee optimizatsii [Urban Environment and Approaches to its Optimization]*, Moscow: Institute of Geography Publications, Soviet Academy of Sciences.

Medvedkov, O. (1979) Modelirovanie gorodskikh makrosistem faktornym analizom [Modelling of Urban Micro-Systems by Factor Analysis Methods], in *Modelirovanie gorodskih sistem [Modelling of Urban Systems]*, Moscow: National Institute of Systemic Studies (VNIISI), pp. 37–40.

Medvedkov, O. (1980, May) Structural Characteristics of Cities as Multifunctional Centers, *Soviet Geography: Review and Translation*, 21: 263–83.

Medvedkov, O. (1980, January) Components in the evolution of urban systems: theory and empirical testing, *Soviet Geography: Review and Translation*, 21: 15–30.

Medvedkov, O. (1980) Modelirovanie prostranstvennykh

157

References

zakonomernostey organizatsii sistem gorodov [Modelling of Spatial Patterns in Systemic Organization of Urban Networks], *Izvestiya Akedemii Nauk SSSR, ser. geogr.,* **4**: 22–36.

Medvedkov, O., and Y. Medvedkov. (1983) Problems of Positionality in City-Regionalization Models, *Soviet Geography: Review and Translation,* **24**: 213–21.

Medvedkov, O. (1988) The Moscow Trust Group: an uncontrolled grass-roots movement in the Soviet Union, *Quarterly Report of Mershon Center,* **12**(4): 27, Colombus: Ohio State University Publications.

Medvedkov, O. (1988) Soviet Cities and their Industrial-Social Performance, *Urban Geography,* **9**: 487–517.

Medvedkov, Y. (1964) O razmerakh gorodov obyedinennykh v sistemu [On the Size of Cities Belonging to a System], in Y. Medvedkov, and Gokman V.M. (eds), *Kolichestvennye metody issledovaniy v economicheskoy geografii [Quantitative Techniques in Economic Geography],* Moscow: Soviet Geographical Society and VINITI, pp. 90–121.

Medvedkov, Y. (1966) Analiz konfiguratsil rasselehiya [Analysis of Settlement Patterns], in *Economico-Geographicheskoye issledovanie kapitalisticheskih raionov mira. Novie tendentsii [Economic-Geographic Research of the Capitalist World Regions. New Trends],* Moscow: VINITI Publications.

Medvedkov, Y. (1977) Settlement in the light of anthropoecosystems Approach, *Geographia Polonica,* **37**: 143–149.

Medvedkov, Y. (1978) *Chelovek i gorodskaya sreda [Human Actor and Urban Environment],* Moscow: Nauka Publishers.

Medvedkov, Y. (1979) Opyt geograficheskogo modelirovaniya gorodskoy sredy [Experience of Geography in Urban Environment Modelling], in *Modelirovanie gorodskih sistem [Modelling of Urban Systems],* Moscow: National Institute of Systemic Studies (VNIISI).

Narkhoz (1987) Narodnoye Khoziaystvo SSSR za 70 let: Yubileynyi statisticheskiy sbornik [Soviet Statistical Year-Book to Commemorate the 70th Anniversary of the Soviet State], Moscow: Finansy i Statistika.

Naseleniye SSSR po dannym Vsesoyuznoy perepisi naseleniya 1979 goda [Population of the USSR According to the 1979 Census (1980) Moscow: Politizdat.

Nelson, H.J. (1955) A service classification of American cities, *Economic Geography,* **31**: 189–210.

Ofer, G. (1987) Soviet economic growth: 1928–1985, *Journal of Economic Literature,* **25**: 1767–1833.

Okabe, A. (1979) An expected Rank-Size Rule, *Regional Science and Urban Economics,* **9**: 21–40.

Okabe, A. (1987) A theoretical relationship between the Rank-Size Rule and Clark's Law of Urban Population Distribution: duality in the Rank-Size Rule, *Regional Science and Urban Economics* **17**: 307–19.

Pallot, J., and Shaw, D.J.B. (1981) *Planning in the Soviet Union,* London: Croom Helm.

Panel on the Soviet Union in the Year 2000 (1987, June) *Soviet*

References

Geography: Review and Translation, **38**(6): 388–433.
Models of City Size in an Urban System (1970) *Papers of the Regional Science Association,* **25**, 221–53.
Pokshishevskiy, V.V. (1978) *Geografiva i naselenie [Geography and Population],* Moscow: Mysl Publishing House.
Polyan, P.M. (1987) Nodal elements of regional support frameworks for settlement, *Soviet Geography: Review and Translation,* **38**: 718–27.
Problemy sovremennoi urbanizatsii [Problems of Modern Urbanization] (1985) Moscow: Soviet Geographical Society.
Rapaway, S., and Baldwin G. (1983) Demographic trends in the Soviet Union: 1950–2000, in *Soviet Economy in the 1980s: Problems and Prospects, Part 2. Selected Papers Submitted to the Joint Economic Committee, Congress of the United States,* Washington, DC: US Government Printing Office, pp. 265–96.
Rashevsky, N. (1943) Contribution to the theory of human relations: outline of a mathematical theory of the size of cities, *Psychometrica,* **8**: 87–90.
Rashevsky, N. (1951) *Mathematical Biology of Social Behavior,* Chicago: Chicago University Press.
Ray, D., and Lones P. (1973) Canonical correlations in geographical analysis, *Geographia Polonica,* **25**: 49–65.
Razmescheniye naseleniya v SSSR: regional nyi aspect dinamiki i politoki narodonaseleniya [Location of Soviet Population: Regional Aspects of Dynamics and Policy] (1986) Moscow: Mysl Publishers.
Richardson, H.W. (1978) Theory of the distribution of city sizes: review and prospects, in L.S. Bourne, and J.W. Simmons (eds), *Systems of Cities: readings on Structure, Growth and Policy,* New York: Oxford University Press.
Rogers, A. (1971) *Matrix Methods in Urban and Regional Analysis,* San Francisco: Holden-Day.
Rom, V. (1986) *Economicheskaya i social naya geographia Sovetskogo Souza [Economic and Social Geography of the Soviet Union],* Moscow: Prosvyeschenie Publishers.
Rosen, K., and M. Resnik (1980) The size distribution of Cities: an examination of the Pareto Law and primacy, *Journal of Urban Economics,* **8**: 165–86.
Sallnow, J. (1987) Belorussia: the demographic transition and the settlement network in the 1980s, *Soviet Geography: Review and Translation,* **28**: 25–33.
Saushkin, Y.G. (1980) *Economic Geography: Theory and Method,* Moscow: Progress Publishers.
Sbornik statisticheskikh materialov: 1987 (Compendium of Statistical Data: 1987] (1988) Moscow: Finansy i statistika.
Shabad, T. (1985) Population trends of Soviet cities, 1970–84, *Soviet Geography: Review and Translation,* **26**: 109–53.
Shabad, T. (1987) Major new Soviet cities, 1985–86, *Soviet Geography: Review and Translation,* **29**: 121–29.
Shatteles, T. (1975) *Sovremennye econometricheskie metody [Modern*

References

Econometric Methods], Moscow: Statistica Publishers.

Smith, D.S. (1979) *Where the Grass is Greener: Geographical Perspectives on Inequality*, London: Croom Helm.

Sovetskaya Geograficheskaya Entsiklopediya; tom 5 [Soviet Geographical Encyclopedia; Vol. 5] (1968) Moscow: Sovetskaya Entsiklopediya.

Thomas, R.W., and Huggett R.J. (1980) *Modelling in Geography: A Mathematical Approach*, Totowa, New Jersey: Barnes & Noble.

Tinbergen, J. (1968) The hierarchy model of the size distribution of centers, *Papers of the Regional Science Association*, **20**: 65–68.

Trus, L. (1977) Class migratsionnyh modeley svyazannyi s sistemoy rasseleniya po Zipfu [A class of Migration Models Related to Zipf-Type Systems of Settlement], in *Izuchenie migratsii [Estimate of Migrations]*, Novosibirsk: Nauka Publishers.

Vapnarskiy, C. (1969) On a Rank-Size distribution of cities: an ecological approach, *Economic Development and Cultural Change*, **17**: 584–95.

Vriser, I. (1980) The Yugoslav settlement system, *in National settlement systems. Topical and National Reports*, Warsaw: IGU Commission on National Settlement Systems, pp. 399–442.

Zipf, G. (1941) *National Unity and Disunity*, Bloomingdale: Indiana University Press.

Zipf, G.K. (1949) *Human Behavior and the Principle of Least Effort*, Cambridge: Addison-Wesley.

Index

Printed in the United States
by Baker & Taylor Publisher Services